GENERATION IN HAZARDOUS AREAS

Ian Staff

ELECTRICAL TRAINING CONSULTANT

First Edition published 2025

2QT Publishing Services

UK

Cover design: Dale Rennard

Images supplied by author

Printed in the UK by IngramSparks Inc UK

ISBN 978-1-7385640-9-5

Introduction / Author

I am a retired Electrical Training Consultant and before my 15 or so years at H.O.T.A. Hull, as a Compex Trainer/Assessor I was 38 years with BP, seven of those years as their Instrument/Electrical Training Officer in charge of all Instrument and Electrical Training and part of their Team in their Training Department where I obtained my Training and Assessing and Training Qualifications.

The Maintenance Electrical Technician in Hazardous Areas

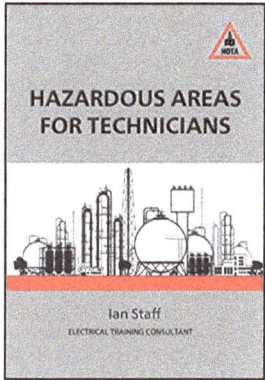

HAZARDOUS AREAS FOR TECHNICIANS Ian Staff ELECTRICAL TRAINING CONSULTANT	**INSPECTIONS IN HAZARDOUS AREAS** Ian Staff ELECTRICAL TRAINING CONSULTANT	**MOTORS & CONTROL IN HAZARDOUS AREAS** Ian Staff ELECTRICAL TRAINING CONSULTANT	**EARTHING & BONDING IN HAZARDOUS AREAS** Ian Staff ELECTRICAL TRAINING CONSULTANT
ISBN 978-1912014958	ISBN 978-1913071615	ISBN 978-1914083013	ISBN 978-1914083112
Batteries & UPS in Hazardous Areas Ian Staff Electrical Training Consultant	**Lamps & Lighting in Hazardous Areas** Ian Staff Electrical Training Consultant	**ELECTRONICS IN HAZARDOUS AREAS** Ian Staff ELECTRICAL TRAINING CONSULTANT	**GENERATION IN HAZARDOUS AREAS** Ian Staff ELECTRICAL TRAINING CONSULTANT
ISBN 978-1914083303	ISBN 978-1914083662	ISBN 978-1914083914	ISBN 978-1738564095

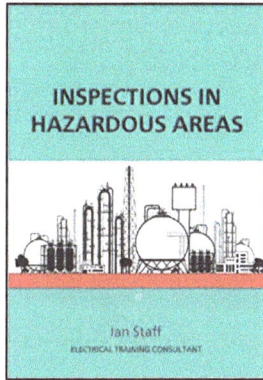

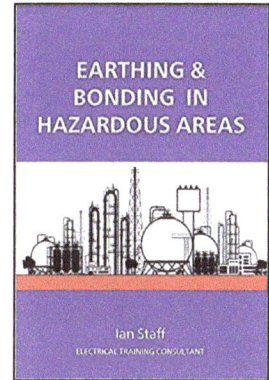

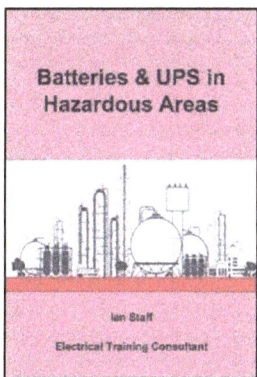

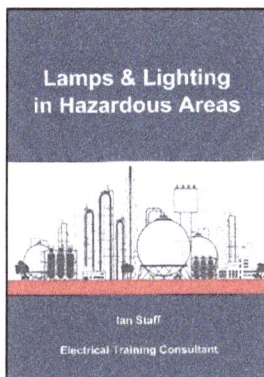

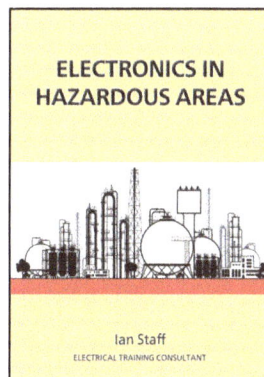

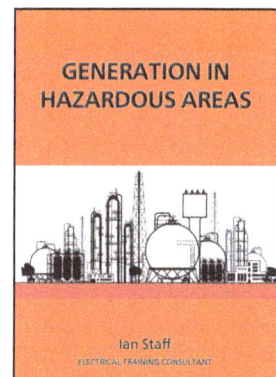

So far I have written seven books, this being the eighth. This set of books shows the knowledge that, in my opinion, is required by an Electrical Technician on a Chemical Factory or a Platform. Having once been there I feel that these books, at a reasonable cost, written at Technician level would demonstrate the full spread of subjects dealt with in everyday life.

The first book talks mainly about Atex, the IEC and EN/IEC Standards, Zones, Gas Groups, Temperature Classes etc. and sets the basis for the other seven: Inspections of Electrical and Instrument Equipment; Motors and Control Circuits; Plant Earthing and Bonding System Types and Testing; Battery Types and Maintenance, including UPS Systems; Lamps and Lighting Systems shows all of the different Lamps/Light Bulbs available, where they are used and controlled in hazardous areas; Electronics in Hazardous Areas would be handy for Electrical and Instrument Technicians alike and this latest one, Generators in Hazardous Areas is on Generators & Alternators some of which may-be of interest, but not actually in your Hazardous Area.

Content

General Information:

This book explains the difference between Generators, Alternators, Dynamos & Magnetos. Although all of these are Generators of sorts they complete their task in different ways. There are also write ups and diagrams of how other different machines work, some of which are included for interest and will not be in this type of Hazardous Area.

My Experience of Generators and Alternators: In my 38 years with BP, I have worked on several different Generators & Alternators, such as a 6.6KV Alternator supplying the factory **HV** Grid along with 3 x 440V Alternators also supplying the factory **LV** Grid. These Alternators were driven by steam turbines with the steam provided in those days by one of our huge coal-fired boiler houses.

I have also had experience on:

1) Brushless Alternators fed by Process Gas Turbo Expanders feeding onto the factory HV Grid electrical system. These were installed in Zoned Hazardous Areas.

2) Induction Generator fed by Process Gas Turbo Expanders feeding onto the factory HV Grid electrical system. These Generators were installed in Zoned Hazardous Areas.

3) Several other smaller Induction Generators fed by Process Gas Turbo Expanders feeding onto the factory grid electrical system. These were installed in Zoned Hazardous Areas.

All the above machines were maintained and tested by our onsite Electrical Department of which I was a part of and carried out training on them for isolation purposes.

Generators & Alternator also to be discussed:

1) Three Phase Generators/Alternators.

2) Induction Generators.

3) Magneto-hydrodynamic Generators.

4) 'Free Piston' Linear Generators.

5) Brushless Alternators.

6) Dynamos.

7) Magnetos.

8) Ward-Leonard System.

9) Integrated Drive Generators.

10) Solar Generators.

11) Floating Solar Farms.

12) Tidal Generators.

13) Wave Generators.

14) Hydrogen Generators.

History: Faraday's Disc Dynamo:

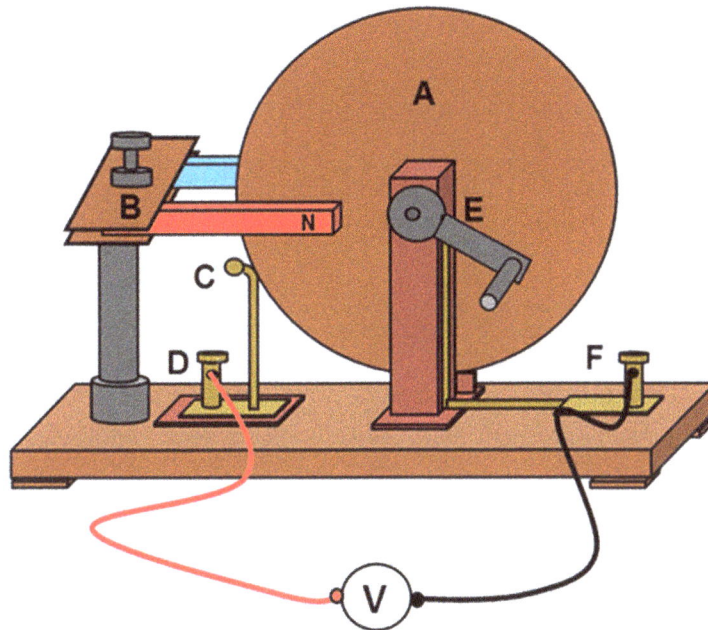

Fig. 1

Michael Faraday (1791-1867): This experiment has been selected from history as it covers so many basic devices like the commutator etc. British Physicist Michael Faraday is responsible for this invention which was one of the first Generator devices in the early 1830s.

Faraday made a device, Fig. 1, which was a copper disc spinning between two poles of a magnet. This device was also called a **Homopolar or Unipolar Generator.** The Faraday Disc Dynamo was quite ingenious for its day.

How does the Homopolar Generator work:

1) It involved a copper disc (A) which spun via a crank inside of a magnetic field created by a magnet (B).

2) Riding on the outside of the disc was an electrode (C) going to wire terminal (D).

3) Wire terminal (F) was connected via a 'riding' electrode to the shaft at the centre of the disc (E).

4) When the disc was turned an EMF (**E**lectro **M**otive **F**orce) appeared between terminals D and F as shown on the voltmeter in the above diagram.

5) It generates by changing the magnetic flux of the magnet when the wheel is revolved at a constant speed.

6) The voltage is very low in the above example, but the current is fairly high mainly because of the very low resistance.

7) The polarity of the EMF depended on the direction of rotation.

8) The EMF produced of course is DC.

History: Heinrich Friedrich Emil Lenz:

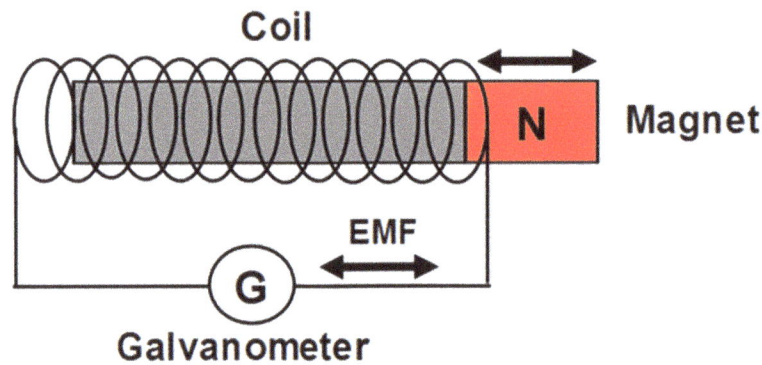

Fig. 2

Heinrich Friedrich Emil Lenz (1804 – 1865): This is another experiment selected from history as it demonstrates very well magnets moving within coils. The start of our basics of generation looks at a Russian Physicist called Heinrich Friedrich Emil Lenz (1804 – 1865). He discovered that if a magnet is moved in and out of a coil, an EMF (**E**lectro **M**otive **F**orce) will be recorded by a galvanometer as in Fig. 2 above.

a) The meter needle will deflect one way as the magnet enters the coil and the other way as it is pulled out of the coil. In fact this was a type of AC (**A**lternating **C**urrent)

b) An EMF will flow from the coil circuit above so long as the magnet is moving, and lines of flux are cutting the coil.

This experiment now proves that if a machine was constructed where there is a coil, and within that coil is a constant moving magnet, then an EMF will flow, and this is called the **'Generator Effect'.** We are actually looking at AC (**A**lternating **C**urrent) and if it was possible to keep the magnet stationary and move the coil then this would have the same effect!

Lenz's Law: The polarity of the induced EMF drives current around a wire loop to always oppose the change in magnetic flux that causes the EMF.

The Potential will increase

a) If the Magnet is moved in and out of the coil faster.

b) If the Magnetic Flux strength is increased.

In the mid 1830s Michael Faraday completed a very similar experiment to that of Emil Lenz using a round bar magnet moving in and out of a cotton insulated coil connected to a galvanometer.

Faraday also changed the solid magnet for an electro-magnet and got the same effect moving it in and out of a coil. Let us remind ourselves about another relevant Law and that is Michael Faraday's Law of Electromagnetic Induction which states that:

A current EMF will be induced in a conductor if it is placed in a continuously changing magnetic field.

Many more examples and diagrams of machines from history can be found in "Motors and Control in Hazardous Areas".

Alternating Current (AC):

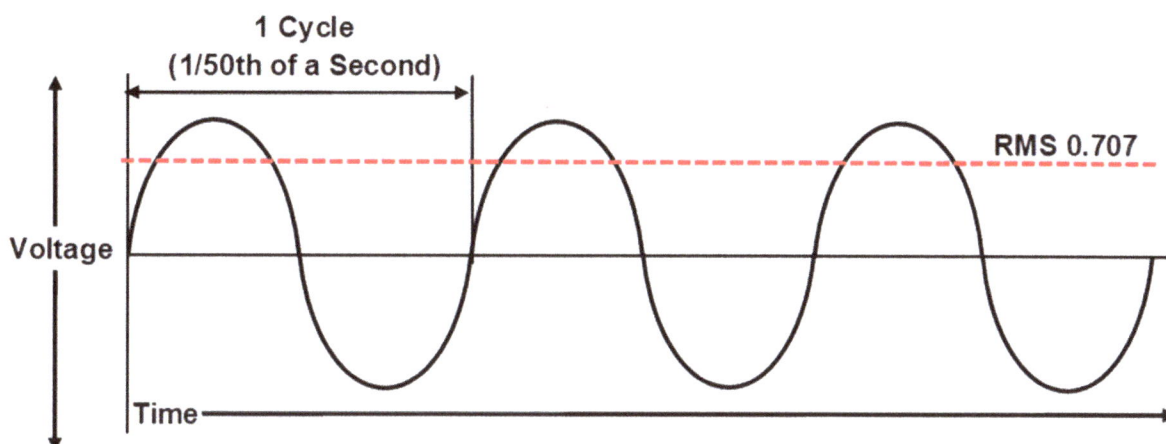

**1 Cycle
(1/50th of a Second)**

RMS 0.707

Voltage

Time

Fig. 3

Alternating current is made up of cycles, as in Fig. 3 above, with each cycle being 1/50th of a second. In the UK these cycles are at the rate of 50 every second. The modern cycles are in the international system of units and are called Hertz, the UK is 50Hz or 50 Cycles per second and the US is 60Hz or 60 Cycles per second.

If we took a UK item of equipment (50Hz) to the US and plugged it into the American system of 60Hz what would be expected, and would it work? The higher frequency would probably work as follows:

a) Motors such as hairdryers and fans will run slightly faster (around 15-20%).

b) Anything which has an Element, it will get hotter.

c) KW of motors will increase with higher frequency.

You would have to ensure that the extra 15-20% did not burn out any of the Elements or Motors or have an adverse effect on whatever they are driving. **There may even be a Fire Risk!**

Now, let us look at the other way round. Would American equipment (60Hz) work on a UK (50Hz) system? Again, with the lower frequency the probable result would be:

a) Motors such as hairdryers & fans will run slightly slower (around 15-20%).

b) Anything with an Element, it will not get so hot.

c) Motors will also lose torque so calculations must be correct!

Sine Waves cannot be very efficient if they are constantly passing through zero so many times a second. The dotted red line in Fig 3 is drawn across the peaks of the sine wave at a certain level, this being 0.707 and this is called the Root Mean Squared (RMS) value and is drawn in this position as being the most efficient part of a AC Sine Wave. *Note: This is not the average of the Sine Wave.*

When instruments are connected to the AC Mains it is this RMS value that they read **NOT** the peak value of the Sine Wave. What is meant by Test Instruments? Well, they can be multimeters, voltmeters or any electricity recording meters. Test instruments, also, only measure **CURRENT** and nothing else, they change the reading to volts, ohms etc. for you on the scale.

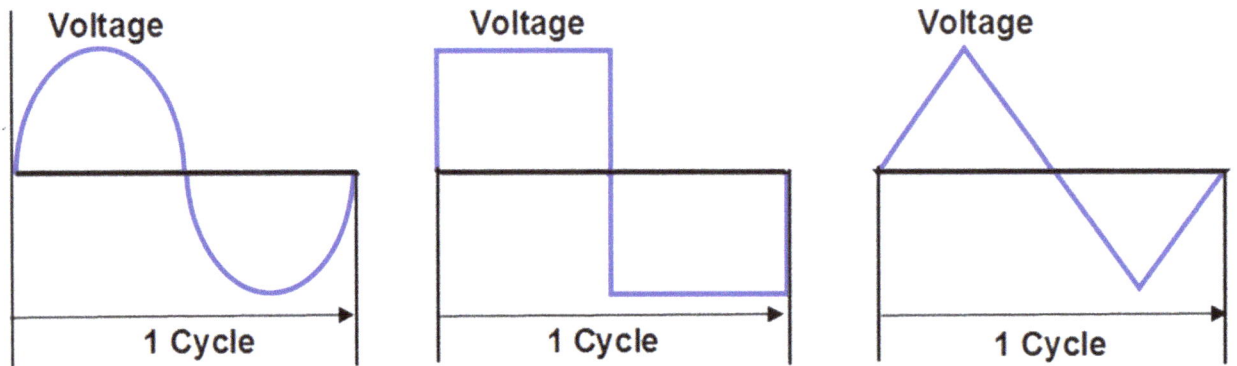

Fig. 4

In each cycle of AC electricity, the Sine Wave would rise to a positive peak and down through zero to a negative peak and back. Sine Waves can be rounded as the common one on the left-hand side of Fig. 4 above, but can be square or even triangular. These could be formed by a Function Generator or Oscilloscope.

If we look at DC for a moment, there is just a positive (+) and a negative (-) and that is all, no matter how large or small the Voltage is. With AC that is quite different as there are 3 phases each 120 degrees out of phase with the other two.

Many industrial items of equipment are three phased including the majority of Electric Motors. Domestic houses use only one of these phases, which is called single phase. The initial installer will spread the three phases over groups of houses so that the load is even.

AC voltage is used in our homes at 200-230 volts (Single Phase) and in industry 400-415 volts (Three Phase). Voltage is transmitted, in very high form around 400,000+ volts, many miles all over the country by the national grid and transformed up and down through substations. AC is much more efficient to transmit, cheaper and more easily controlled than DC.

AC Voltage is easily stepped up or down by equipment called a Transformer, this would present all sorts of problems transmitting DC at an equivalent high voltage. DC higher voltages can be stepped up and down by a piece of equipment called a Boost Converter.

When we talk of imports, we usually think of goods, but we also import electricity from countries like Norway and France. The longest cable run in history is around 450 miles and runs from Norway, where hydro (water) power is used, to Northumberland in the UK and it can power 1.4 million homes.

If DC was used for the cable run from Norway, the logistics would have been extremely complex, not counting the volt drop which could have occurred. In my opinion, we must be careful when importing power as governments and policies change and the electricity that we are importing could be cut off leaving the UK with a shortfall.

Wiring colours have now come into line with Europe. Three phase (400-415V) colours used to be red, yellow & blue with green & yellow for earth, but now the colours are brown, black and grey with green & yellow for earth. Wiring core colours for domestic where there is, usually only single phase, the wiring colours are brown for live, blue for neutral with green & yellow for earth. *Note: Twin and earth wiring, in the past used to be red for live, black for neutral and green for earth. Red & black cores today are the colours of DC.*

High voltage underground cable sheaths are red, but the cores must be colour coded by the electricity companies as they go. There are protection boards over the top of the underground cable with a yellow tape label explaining: 'Caution: HV cable below'

Do Generators & Alternators generate AC? Alternators, because of their design, generate AC only, but Generators can generate AC or DC by going through either a Commutator or Spinning Diodes.

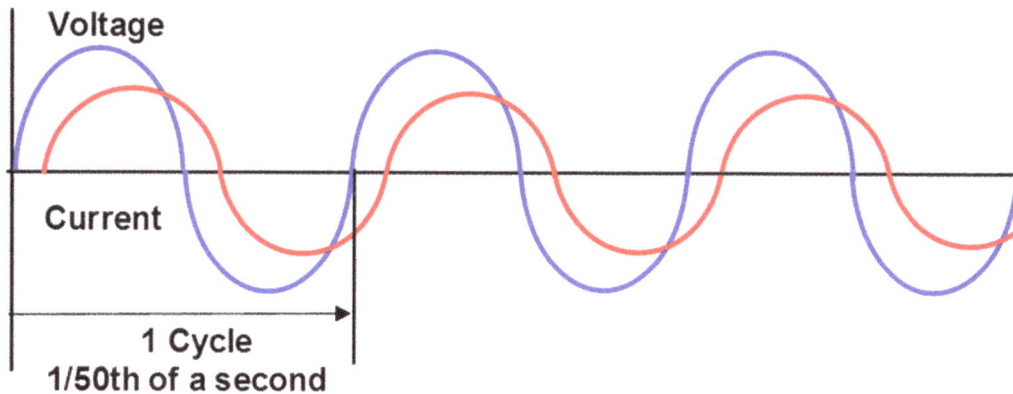

Fig. 5

Back to AC. If the sine wave is voltage, what has happened to the current? Well, looking at Fig. 5, the blue sine wave is voltage and red sine wave is current, and the red sine wave is lagging the voltage, which indicates that this is an **'Inductive'** load such as a Coil. The electric supply company looking down the power line at a large factory with the amount of Motors, Transformers etc., that are just coils, then a lagging current would be the most common.

Do we need to change the lagging current to bring it more in line with the voltage? If we did that the current would be said to be moving towards **Unity.** The cosine **(Cos φ Phi)** of the angle by which the current lags or leads the voltage, as in Fig. 5, is what is called the **'Power Factor'** which suppliers try to keep to as near **'Unity' (1)** as possible. If the Power Factor was to drift then the current could increase to the point where cables started to get hot even if, in theory, they are the correct size.

Looking inside a fluorescent fitting they have a small capacitor, and this is for Power Factor Correction. Because of the large coil, with many windings, in the fluorescent lamp unit i.e. the Choke, each light unit would have its own capacitor as a Power Factor Corrector which in turn assists the main factory Power Factor, taking into consideration the large amounts of fluorescent lighting installed.

Motors and Transformers are compiled of copper coils and **DO NOT** have built in Power Factor Correction. If the factory is quite large and there are many Motors and Transformers then, as mentioned above, steps may have to be taken to correct the Power Factor such as Capacitor Banks.

Unity is '1' so here, for a lagging Power Factor, we have learned that installing capacitors will cause a change in Power Factor and nudge it towards Unity. When we look at the formula of true power i.e. the power that we use and the electricity provider charges us for, it is Watts = Volts x Amps x Power Factor. If we therefore have the Power Factor at Unity i.e. '1', the formula becomes Watts = Volts x Amps x 1. If the Power Factor was not Unity and it was, say, 0.5 then our Formula would be Watts = Volts x Amps x 0.5 which ends up very poor.

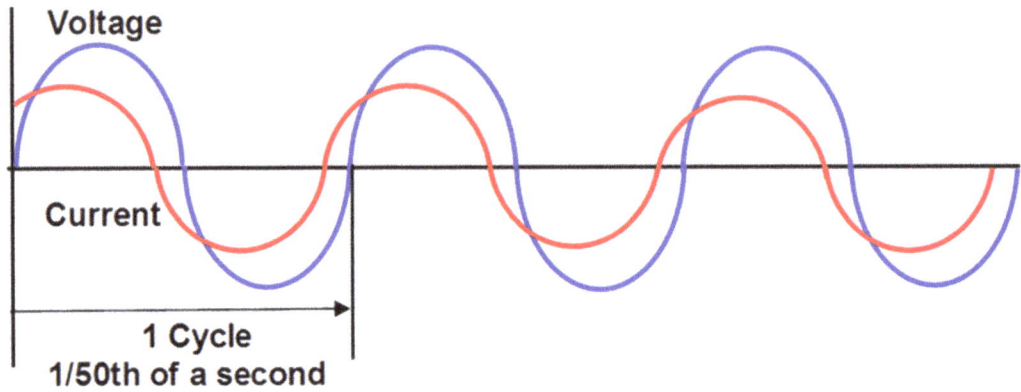

Voltage

Current

1 Cycle
1/50th of a second

Fig. 6

Fig. 6 above is very similar to Fig.5 in that the blue sine wave is voltage and red sine wave is current. However, this time it shows that the red sine wave, the current, is leading the voltage which would indicate that this is a 'Capacitive' load. For a factory to have a mostly Capacitive total load would be unusual and to adjust the current towards **Unity (1)** here, they would have to introduce inductance such as coils.

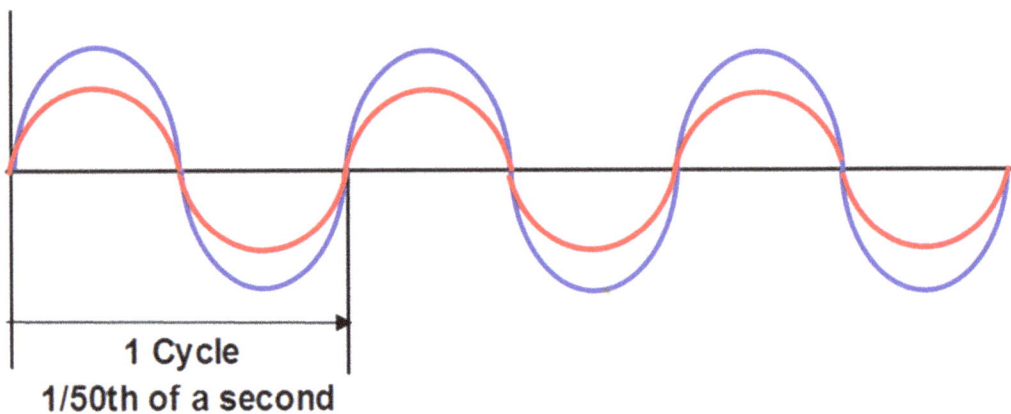

1 Cycle
1/50th of a second

Fig. 7

To recap: If there was an 'Inductive' load then the current would 'lag' the voltage and in a 'Capacitive' load then the current would 'lead' the voltage. In an ideal world the current would be directly in phase with the voltage which would, of course, be 'Unity' (1) as in Fig.7 above.

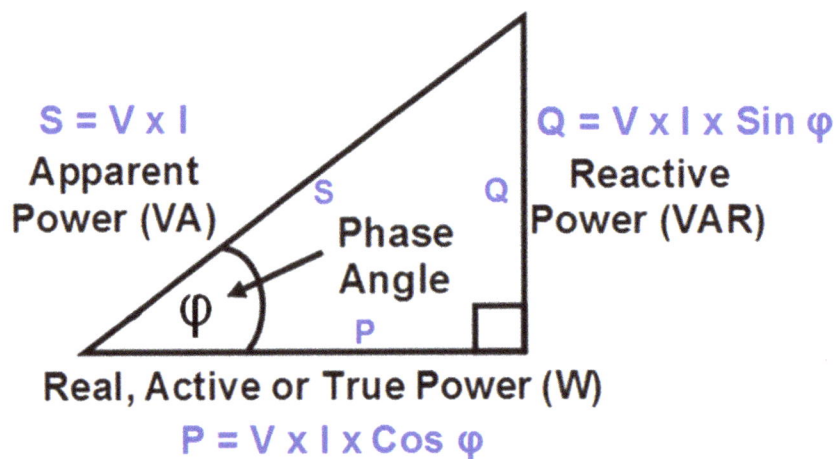

$S = V \times I$
Apparent
Power (VA)

$Q = V \times I \times Sin\ \varphi$
Reactive
Power (VAR)

Phase
Angle

Real, Active or True Power (W)
$P = V \times I \times Cos\ \varphi$

Fig. 8

Fig 8 above shows the 'Power Triangle' with the 'Phase Angle' between Real/Active/True Power in Watts (W) and **Apparent Power** in Volt Amps (VA). To obtain the Power Factor here, it would be Real/Active/True Power (Watts) divided by Apparent Power (Volt Amps).

When we talk about Real/Active/True Power, this is what you pay the Electricity provider for, but Apparent Power is the total power in the circuit and is the combination of volts and amps, so is measured in 'Volt Amps'. This value ensures that all components in a particular circuit have the capacity to accept the load. Transformer and Generator capacity **(Electrical Size)** will be measured in **Volt Amps** (VA or KVA).

Reactive Power is slightly different as here a Capacitor or, for that matter, an Inductor can have energy stored within the device. This energy is **'Reactive'** power and is measured in **Volt Amp Reactive** (VAR). This power can be released as the circuit is working and is measured in VARs.

So how else can the Power Factor be corrected and altered towards Unity? The Power Factor can be corrected by inserting huge Generators, Synchronous Motors, Inductors or Capacitors into the system. Industrial Power Factors may run at around **0.8** but the supply contractor will sometimes indicate what Power Factor they would like a company to run at. So how do we get to this 0.8?

Induction Motors as their name might suggest are **ONLY** Inductors i.e. coils, and they will cause a **'Lagging'** Power Factor where the current lags behind the voltage. Large factories will have many three-phase Induction Motors on their complexes and looking down the line from the supplier this adds up to one huge coil.

Synchronous Motors are a different motor altogether from the Induction Motor and can be used for Power Factor Correction. The Changing Power Factor is achieved on a Synchronous Motor by varying the Field Current where an increase in Motor Field Current can:

1) Cause a bad Power Factor say **0.1– 0.6.**

2) Cause the Power Factor to Lag (**Inductive**)

3) Cause the Power Factor to achieve **Unity (1)**

4) Cause the Power Factor to Lead (**Capacitive**)

Generator theory is slightly different but can still alter the Power Factor. Some Generators operate with low Power Factors. The 'Reactive Power' generated, which has been touched on **(VARs)**, will go up as a result, to meet the output resulting in a higher current. The way to correct the Power Factor in this case is to add Inductors or Capacitors to the circuit. This can also be used to have an effect on the overall Power Factor of the Factory.

Resistance (R) is the opposition to current flow in a circuit and measured in Ohms (Ω). This could be a physical resistor, cable, equipment etc. Impedance (Z) is the opposition to a current flow with voltage (Volts) applied in a circuit. It includes both 'Resistance' and 'Reactance' and like Resistance, is measured in Ohms (Ω).

Hi Fi Speakers, are measured in **'Impedance'** which, in theory, can only be measured when they are working. The test on a Socket System would be an Earth Loop 'Impedance' Test as this would include the installation i.e. sockets, cable, distribution board and supply transformer coil feeding the system. Having a coil in a circuit would be measured in Impedance.

Inductive Reactance mainly applies to the resistance to changes to AC current in an inductor such as a coil in an AC circuit. Usually symbolised by **'X_L'** its end objective is a **Phase Shift** between voltage and current.

Generators, like electric motors, deal with magnetic fields. Comparing a magnetic circuit to an electrical one sometimes demonstrates the best explanation. **Magnetic Reluctance** is the opposition to magnetic flux within a magnetic circuit, it could be called 'Magnetic Resistance' similar to Electrical Resistance. Generator and Alternator Rotors & Armatures are subject to Magnetic Reluctance where the measurement is in Ampere Turns per Weber-Inverse Henry.

Later in the book we talk about Generators requiring **Residual Magnetism** to enable them to generate. Residual Magnetism relies on a phenomena called **'Retentivity'** which is the ability of a material, such as iron, to retain a certain amount of magnetism from when it last ran, hence without Retentivity, or Residual Magnetism, then Generators will run with nothing seemingly wrong but the output would be zero.

Like Resistance in an Electrical Circuit, **'Permeability'** is how easy it is for a material to form a Magnetic Circuit by allowing Magnetic Flux Lines to move through it. Ferromagnetic Materials are materials that can be strongly attracted to a magnet and are easily made into Electromagnets. Iron is a prime example.

Electromagnets are made up of these materials surrounded by coils and only become magnetised if an AC or DC current flows through the coil, becoming demagnetised if the current is switched off. Armatures are a prime example, and it is important that the Core Material used does not remain magnetic when the power is removed. A **B-Field** is another name for a Magnetic Field. The next question is, can we make the magnetic field of an electromagnet stronger? Magnetic Field Intensity or **Flux Density** is how strong the Magnetic Field actually is and Alternators rely on Flux Density from their Exciter.

Direct Current (DC):

DC Voltage is the most efficient when comparing it with an AC voltage. One of the main drawbacks, as mentioned earlier, is that it is not efficient during transmission. DC cannot easily be transmitted over large distances like the National Grid, as transformers etc. obviously do not work on DC as they require an oscillating frequency!

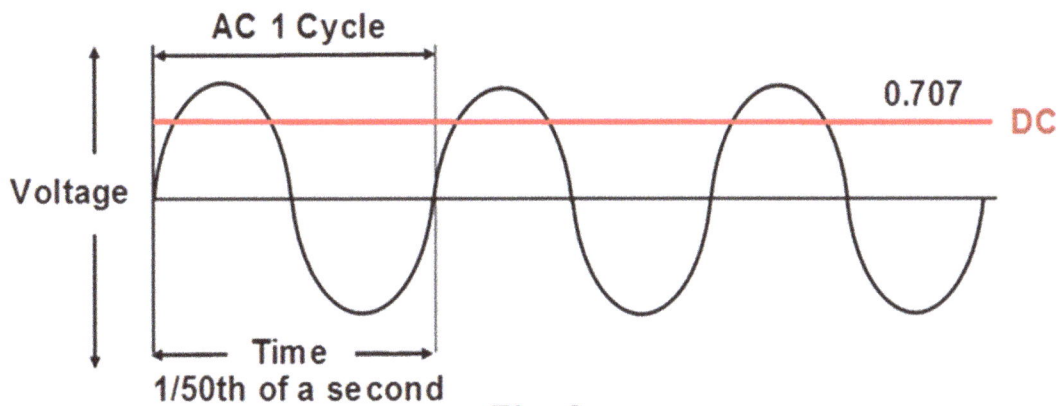

Fig. 9

If we start with AC and want to change that to DC then this is fairly easy. It is commonly done with car battery chargers with Diodes & Rectifiers. AC Sine Waves are shown in Fig. 9 above and the DC is shown, in red, which I have set at the RMS value of 0.707 of the Sine Wave. So, if DC is the most efficient voltage, how is it obtained? There are several types of DC and examples of these are shown on the following pages.

If we take a battery for instance, the output of which would be a DC straight line similar to the red line in Fig. 10 below, there could not be a better example to show DC. The Voltage would depend on the Battery Output. DC however, can be created another way, say, from AC and that would be a little bit more involved, so how is it done?

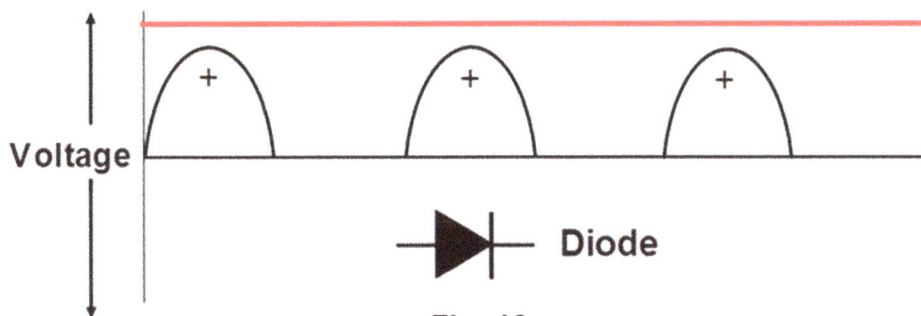

Fig. 10

The AC Sine Wave shown previously in Fig. 9 has a Silicon Diode inserted into the supply as per Fig. 10 above. As the AC Voltage goes first positive and then drops down to negative and then positive again, inserting the diode would stop the Sine Wave from going negative, hence there would be a type of half wave DC rectification, be it very electrically **'noisy'**. The diagram in Fig. 10 would be a **positive Half Wave Rectification**, there can of course be a negative Half Wave Rectification.

What is Electrical Noise? Well, just installing a single Diode would leave gaps in the distance between the positive peaks (half Sine Waves) so there would be a peak every 1/50th of a second and then a gap before the second peak etc. So appliances, in theory, would go off between the peaks. If this was to be in, say, certain instrument circuits then any readings could become very distorted. It is used quite a lot in electric welding where there does not necessarily have to be a steady voltage.

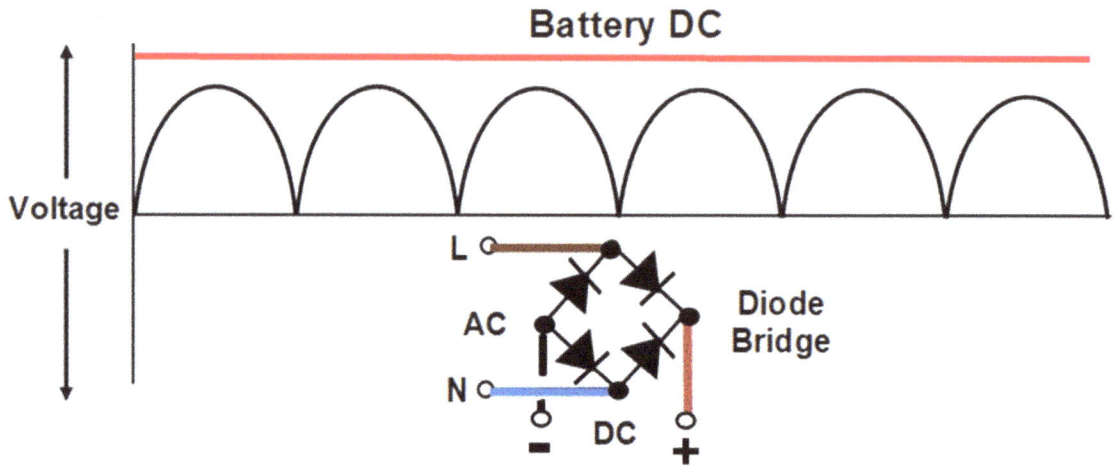

Fig. 11

Fig. 11 above shows AC to DC using a Diode Bridge. It was invented by Karol Pollak in 1896. In fact, this bridge would invert the negative part of the Sine Wave up to the positive to fill in the gaps which would result it being less 'noisy'. Although the system is getting nearer to the DC straight line, it can still be improved. This procedure is called **Full Wave Rectification** or in some descriptions a 'pulsing DC'. This procedure can also be carried out with two Diodes and a Centre Tapped Transformer, although that would incur more expense.

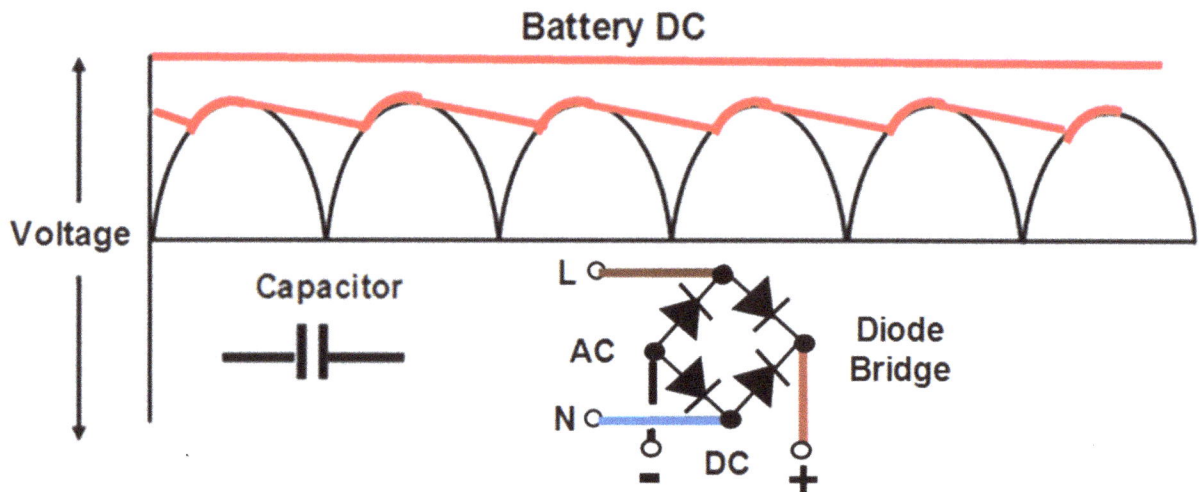

Fig. 12

To make the DC even smoother, we can add a Capacitor to the Full Wave Circuit. This would fill in the **'valleys'** between the peaks: it may be as good as can be achieved artificially. This DC would be quite smooth and could be used for sensitive equipment such as specialist instruments and electronic circuits where smooth DC is required.

DC is very similar to Single Phase AC and with DC there is no such thing as Phases for instance in AC there are 3 Phase systems, but instead of a Live and Neutral, the cores are called Positive and Negative. So the current here **COULD** be described as always Unity (1) and always in phase with the voltage, so no Power Factor is ever involved.

It has been shown up to now that changing AC to DC is quite easy, but to change DC to AC is a little more complex and a piece of equipment called an **'Inverter'** is required. **Note here** three phase AC can be formed from DC. Equipment can be obtained where single-phase AC is fed in, changed to a stable DC and so a three phase AC power is created from the DC. This is called a Single Phase to Three Phase **'Converter'** and is AC & DC combined.

16

Many ask, "do Generators generate DC?". Generators **DO NOT** actually generate AC, and this is changed to DC by a Commutator or Spinning Diodes on the Armature. The DC can be stepped down using equipment such as a **'Rheostat'** or a **'Variable Resistor'** and can be stepped up using a device called a **'Boost Converter'**. Connecting several batteries together in series will also increase the voltage.

Will AC Lighting work on DC? **SOME** AC Lighting would work on DC, but certain facts about the Lamps need consideration:

a) **AC Incandescent Lamps** are those with an Element, and should work okay on DC but maybe less efficiently. Torch bulbs do work on DC all the time.

b) **AC Discharge Lamps** are those which require Ballasts and Chokes, so **NO** they would not work on DC as these devices control current and require oscillating voltages.

AC equipment will **NOT** usually function when plugged into DC as there is no **'Inductive Reactance'** so, the DC may see the AC equipment with its coils as a short circuit and burn out the equipment due to the coil not being able to stand the high DC current. It is possible to obtain **'Universal'** equipment such as a Universal Series Motor in, say, an electric drill or certain fans for instance.

So, "will DC equipment work on AC?". The answer is again **NO.** Electronics, such as Transistors and certain Capacitors, in the equipment will usually burn out. Again, any equipment with a Universal Series Motor should run okay.

Static electricity is DC and includes lightning. It can be a very high voltage and a huge nuisance. Static usually builds up on an insulator and can be caused by natural phenomena such as wind. Another type of electricity is Triboelectricity. Have you ever got out of the car, touched the car frame and received a shock? Well, this is called Triboelectricity, where electrons collect on insulated clothing, such as Nylon. It gathers electrons when rubbing on other materials such as leather which will freely give up electrons and this DC will discharge to metal. Static is a consequence of Triboelectricity.

DC is 'Unidirectional' as flows in one direction only. To change direction, a change of polarity from Positive to Negative on the power source will suffice. A mention here to the findings of a man called Benjamin Franklin who discovered that, Current flows from Positive to Negative and Electrons flow from Negative to Positive. We cannot see current or electrons, and so it will not cause us too much concern. However, some DC equipment will be polarity sensitive so ensure that the DC connections are correct before power is put on otherwise severe damage can occur.

Can DC Voltage be measured? **YES** and **NO.** Let me just be clear in this case, **ALL** instruments measure **CURRENT** and change to Ohms, Voltage or anything else on the scale for you. If you take a multi-meter and switch to **Current,** carry out your readings and then switch to **Voltage** you have not changed the instrument here, all you change is the impedance. To measure DC when you change the scale it just adds a diode.

DC Cable Colours **are** Red for Positive (+) & black for Negative (-). Around the 1960s cable core colours on AC twin and earth cable were also red and black. Other colours which may be used on systems such as **D**istributed **C**ontrol **S**ystems (DCS), may be other colours such as, orange and white, as may cables from other countries.

DC is used a lot in the Motor Vehicle Industry, although some electric vehicles these days may be fitted with an inverter to AC. Crane Motors will be DC because of the extra control they have which is better than AC. Toys such as train sets and torches are DC. Electronics and Solar Panels will also be DC.

Are vehicles Positive or Negative Earth? In most situations with vehicles, there will be a straight two wire system, the metal case of the vehicle being connected to one battery terminal, usually the negative (-), which makes the vehicle case one electrode. Before 1968, vehicles were positive earth, and anything connected into the vehicle was polarity conscious. After 1968, vehicles became negative earth, causing the case of the vehicle to be negative (-) resulting in a **'Cathodic Protection'** which cuts down corrosion, (Anodes (+) corrode). How many vehicles do you see these days rotting away?

Do DC Systems require Earthing? This is not an easy question to answer. Certain manufacturers designed DC equipment requiring a three-wire system with a third wire, earth, connected to the earth terminal. Manufacturers have to look at a safety problem of a core coming off the terminal and touching the case of the equipment causing all sorts of problems including tripping supplies. Look at the manufacturer's **'Hook Up'** or if Intrinsic Safety, the company **'Loop Diagram'** and see what is recommended in the design as these diagrams should be 100% correct.

Screens are required by some instruments, especially Intrinsic Safety, a 'Screen' wire connected to a high integrity earth of 1Ω, but this will be indicated on the Loop Diagram. The Screen is a silver insulating foil that surrounds the cores in an Intrinsically Safe Cable (Sheath Blue) **NOT** the bare wire coming out of the end. It is actually a Faraday Cage around the cores to stop invading induced currents and the screen wire is in contact with the screen all the way through the cable.

Screen connection cores are identified from Electrical Earths by colouring the IS cable, bare Screen core with sleeving. Screen earth connection cores will usually have **ALL GREEN** or **ALL WHITE** sleeves as against green and yellow for Electrical Earths. Companies can use their own colours provided they are clear and marked on their Loop Diagram.

There are two main Earthing Systems **'Insulated'** Earth and **'Uninsulated'** Earth. Screens should be connected to. **'Insulated'** Earth, which is 1Ω, so named because they are insulated right up to their Earth Rod and can also be called **'Instrument'**, **'Clean'** or **'IS'** Earth. Electrical Earths which are called **'Uninsulated'** or **'Dirty'** Earths because are not required to be insulated up to their Earth Rod and can be up to 5Ω. Lightning Conductor Earths do not have to be high integrity earths, so long as they do go to ground and can be up to 10Ω.

AC Alternators & AC/DC Generators:

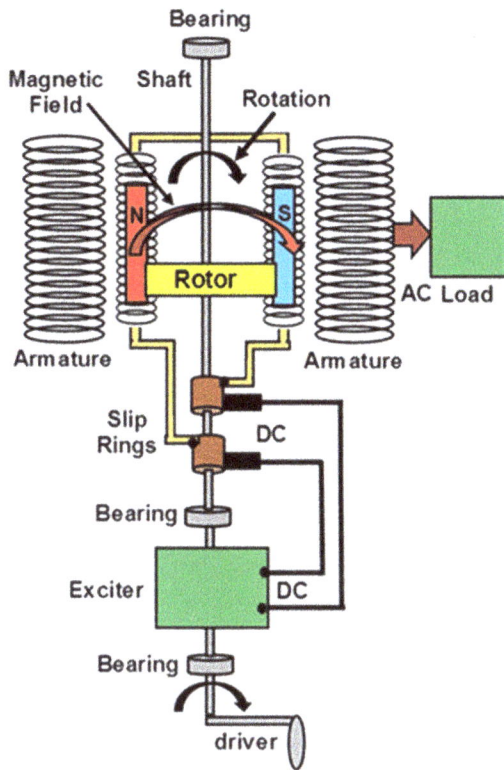

AC Alternator

Fig. 13

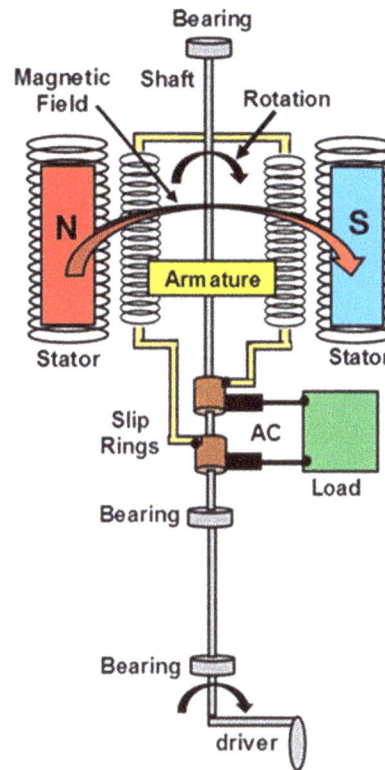

AC Generator

Fig. 14

We will leave Dynamos and Magnetos to one side for the moment and look at AC Generators and AC Alternators and how they differ. Before we look at the difference between an Alternator and a Generator, just remember Heinrich Friedrich Emil Lenz and his experiment of moving Magnets within Coils or moving Coils outside of Magnets.

Alternators which are sometimes called **Synchronous Generators,** as in Fig. 13, **ONLY** generate AC. The AC Generator diagram in Fig. 14 may look similar to Fig. 13, but the way they operate is quite different as will be explained as we go.

An Alternator is a spinning Electromagnetic Rotor (Magnet) in outer Field Coils. To make a Rotor Spinning Coil into an Electromagnet it has to have power. This is achieved by fitting rotating Slip Rings and Brushes onto the shaft to receive the DC current from another Generator called an **Exciter.** The Load here is taken from the Rotating Magnetic Field in the Stator Field Coils which in this case are called a **Stationary 'Armature'**. Alternators can be Self Excited or Separately Excited as we will discuss later.

Generators generate AC as in Fig. 14. A Generator is a spinning Rotor Coil or **'Armature',** as it is called, spinning within an Electromagnet provided by the Field Coils of the **Self Excited** Generator. The AC Load is taken, in this case, off the Slip Rings. Generators are excited by **Residual Magnetism** from the last time they ran.

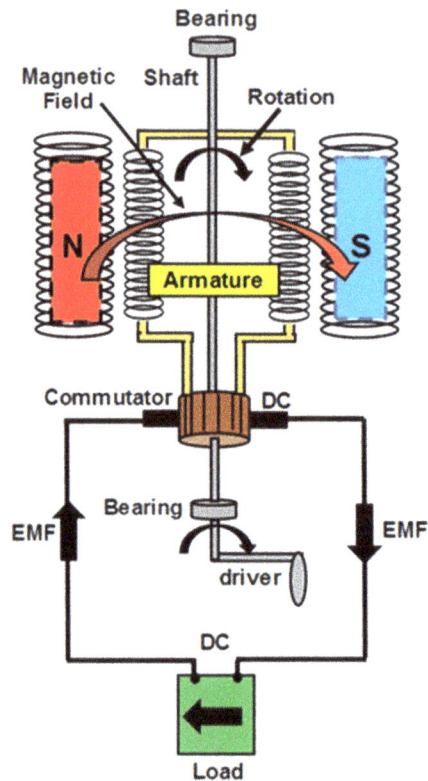

DC Generator

Fig. 15

We have said that Generators generate AC, unlike the Alternator that can be changed to DC by fitting two devices called a Commutator and Brushes, which are shown in Fig. 15 above. Another option is to fit Spinning Diodes which are discussed in the Brushless Alternator section later. The Commutator above has several segments.

To recap then; with a Generator there is a Coil called an Armature spinning in a Magnetic Field (Electromagnet) produced by the Self Excited Field Coils. The DC is delivered to the load by courtesy of a Commutator and Brushes. A DC Generator will run quite happily as a DC Motor.

We have mentioned that Alternators generate their power by a Rotor Magnet spinning in Field Coils. But what gives that Rotor Magnet the power to become a magnet? The answer is an Exciter, and this feeds DC power onto the Alternator Rotor by means of Slip Rings and Brushes. Exciters are DC Generators complete with Commutators & Brushes or Spinning Diodes. The Exciter is usually on the same shaft as the Generator so they both run at the same speed, but Alternators can have separate Excitation, as shown later, although this is not the norm.

Automatic Voltage Regulators (AVRs) control the Exciter Field and thus the Alternator output, by regulating the current to the Alternator Rotor Magnet. So, what are AVRs and how do they work? The simple answer, without going too deep, is that it is a piece of equipment in an Alternator control circuit that keeps the Output Voltage at a particular pre-set value even if the load was not stable and prevents surge. As well as being called an AVR, this unit can also be known as an Automatic Voltage Stabiliser (AVS) because stabilising is what it actually does.

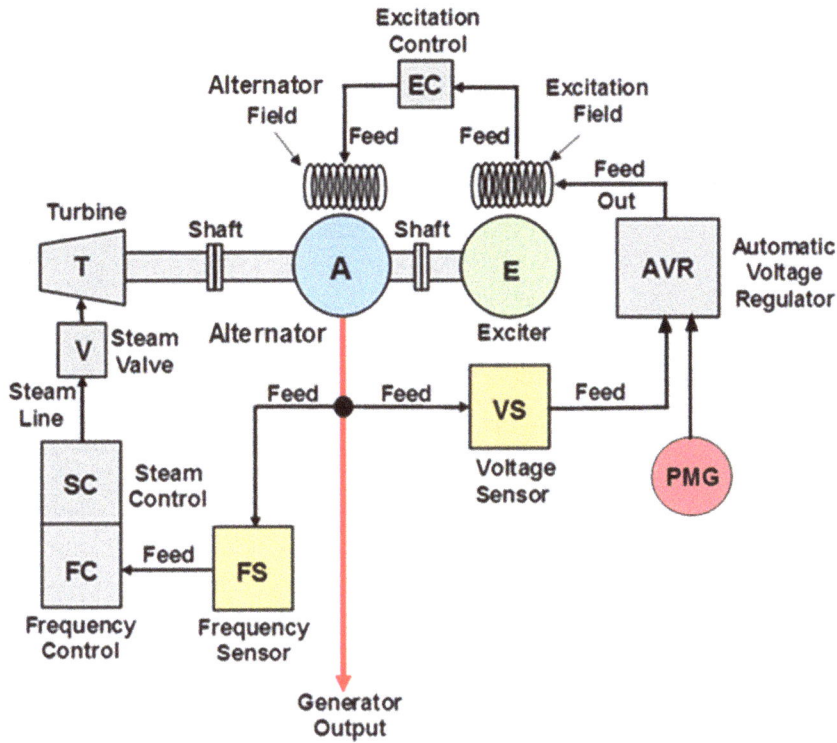

Fig. 16

The AVR as you can see in Fig. 16 is fed information from the voltage sensor as to an over or under voltage on the Alternator Output, and the AVR will adjust the excitation from the exciter to the Alternator Main Field Flux Density accordingly. Higher or Lower Voltage depends upon Winding Turns inside of the Alternator or the density of the Magnetic Flux which is where the Exciter comes in.

With no AVR there could be problems if the load on the Alternator was to fluctuate or, for some reason, more voltage is required, and the Alternator could **'Hunt'** as it tries to find a stable voltage. Also, there could be problems with the machine stabilising if the voltage on the grid that it was feeding onto was to fluctuate. The Permanent Magnet Generator (PMG) shown in Fig. 16 is where the AVR gets its own power from and could run via a small Gearbox off the main Alternator Shaft.

A Turbine is usually the Driver, so inevitably this machine governs the speed of the Alternator. The Rotor Shaft Speed must be kept fairly stable, as any change will alter the frequency and by increasing can also lead to higher voltage although this might be the more difficult to achieve.

A Frequency Sensor monitors the frequency on the Alternator Output feeding whatever drives the Turbine Valve to adjust the Speed. Six different Alternator Synchronous Speeds are listed below for 50HZ in the UK and 60HZ in the US. The US speeds are slightly faster with the higher frequency.

  c) 2 Pole machine speed   50 HZ 3000 RPM   60 HZ 3600 RPM

  d) 4 Pole machine speed   50 HZ 1500 RPM   60 HZ 1800 RPM

  e) 6 Pole machine speed   50 HZ 1000 RPM   60 HZ 1200 RPM

  f) 8 Pole machine speed   50 HZ 750 RPM   60 HZ 900 RPM

  g) 10 Pole machine speed   50 HZ 600 RPM   60 HZ 720 RPM

  h) 12 Pole machine speed   50 HZ 500 RPM   60 HZ 600 RPM

As can be seen, Alternators, as Electric Motors, decrease in speed as the number of Poles increases.

The formula for working out the speed of the Alternator is:

Ns (Synchronous Speed) = 120 x f (Frequency) divided by P (Number of Poles)

As an example, Ns (Synchronous Speed) = 120 x 50 (Frequency) = 6000

Divided by P (Number of Poles) which is 6000 divided by, say, 8 (Poles) = 750 RPM.

Check the figures that are arrived at with the figure on the list above for an 8-pole machine and try some figures of your own. Remember, if this is done on American figures then the Frequency will be 60 Hz, and the speed will be slightly higher.

Alternator over-speeding is not so common with the AVR Unit looking after the output, but what might cause the Alternator to over-speed? Well, the Load Circuit Breaker trips on load with the Primary Driver still connected. It is possible in this instance for the Alternator to over-speed, so if this is deemed a catastrophic problem for the Engineer, then steps must be taken for a system shutdown and the time period to be looked into.

Faulty AVRs can also cause an over-speed. It is always advisable especially in the case of for example, a Power Station with several Alternators, to have a spare AVR set up ready for replacement. AVR faults can cause too much excitation and as well as causing the Alternator to not only speed up but severely overheat.

Faulty PMGs or the link to the AVR could also be a problem as they provide power for the AVR itself.

Note: The AVR could receive its power externally without a PMG.

Faulty Voltage or Frequency Sensors could cause problems with the Alternator speed and so it is advisable to have spare sensors ready to replace any faulty ones. There is one fault that could cause over-speed and that is the Turbine Driver, but that is not a problem with the Alternator itself.

Let us now move from an Alternator to a Generator. Residual Magnetism must be present for the Generator to start generating. If this Residual Magnetism is not there from when the machine last ran, then there will be no output.

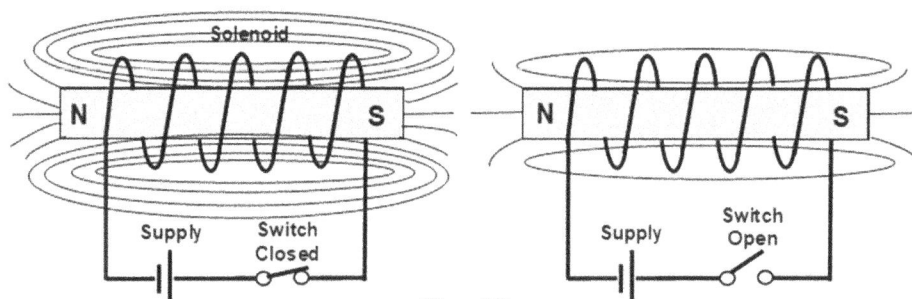

Fig. 17

The Residual Magnetism is a few lines of Magnetic Flux lingering around the Poles from when the Generator was last run (Fig. 17 Right). When the Generator starts the Armature Coils cut these Residual Lines of Force and initiate self-excitation as the current starts to flow. Without any Residual Magnetism the Generator would be driven by the Driver, but there will be no output even though everything appears to be okay. A domestic generator could suffer the same fate.

Winding Configurations depend upon the type and design of the Generator. Not all Winding Configurations will be the same. It would be less problematic if we could use Permanent Magnets, but for practicality and constant efficiency, that is not possible.

DC Generators come in several different configurations.

  a) Series Wound.

  b) Shunt Wound.

  c) Compound Wound Short Shunt.

  d) Compound Wound Long Shunt.

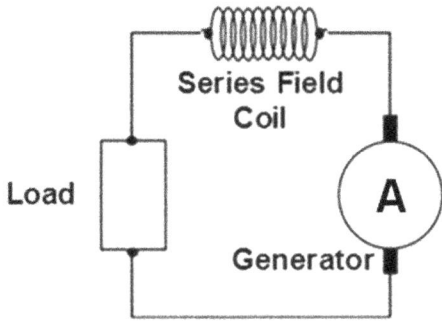

Fig. 18

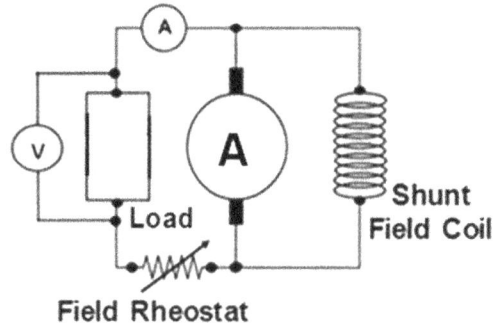

Fig. 19

The **Series** winding configuration as in Fig. 18, where the Field Coil winding is in **'Series'** with the Armature High Resistance. This design is a configuration that is not used much in present day. For self-excitation, it relies on Residual Magnetism but first it must be connected to the load. There are less turns of wire in the Field Winding than the **'Shunt'** configuration. The **Output voltage** is **'Directly'** proportional (as one **increases**, the other **increases** at the same rate) to the **Load Current.**

The **Shunt** winding configuration, Fig. 19, where the Field Coil winding is in **'Parallel'** with the Armature. Thick wiring with fewer turns and Low Resistance as was in Fig. 18. For self-excitation, again it relies on Residual Magnetism, but no permanent magnets. The **Output voltage** is **'Inversely'** proportional (as one **increases**, the other **decreases** at the same rate) to the **Load Current.**

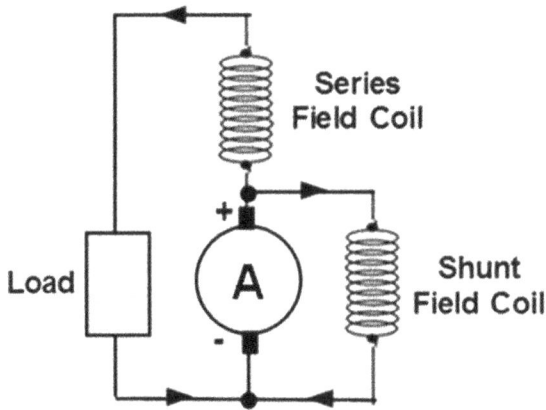

Fig. 20

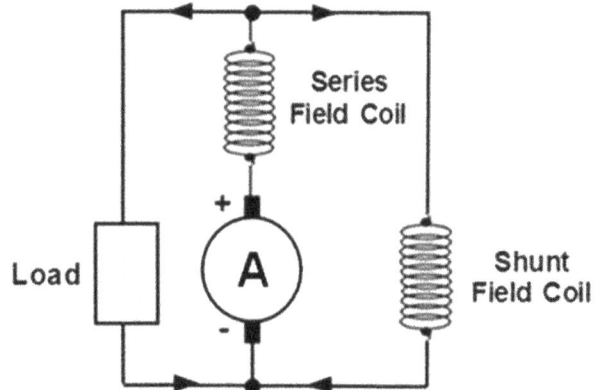

Fig. 21

The **Compound** Wound Generators are different to what has been done up to now with the independent Series and Shunt Windings. Here the two different Field Windings i.e. Shunt Winding and Series Winding, are both on the same Pole. Compound Wound DC Generators can further be divided into two more categories namely:

  1) **Short Shunt** Wound Fig. 20.

  2) **Long Shunt** Wound Fig. 21.

23

There are disadvantages between the Series Wound and Shunt Wound DC Generators, and the Compound Long and Short Shunt Wound overcomes them all.

Compound **Short** Shunt winding configuration shown in Fig. 20 is as follows:

1) The **'Series'** Field Current is **'Equal'** to the **'Load'** Current.

2) The **'Shunt'** Field Coil is connected in parallel to the Armature.

3) Self-Excitation relies on Residual Magnetism.

Compound **Long** Shunt winding configuration shown in Fig. 21 is as follows:

1) The **'Series'** Field Current is **'Equal'** to the **'Armature'** Current.

2) The **'Shunt'** Field Coil is connected in Parallel to the Armature and Series Coil.

3) Self-Excitation relies again on Residual Magnetism.

Compound Windings can be further broken down into:

1) **'Cumulative'** Wound and

2) **'Differential'** Wound.

The two Compound windings (Shunt & Series) are wound on the same Pole. The Shunt Field will be stronger than the Series Field.

So they could be:

1) **Cumulative:** if the 'Series' Winding Flux **assists** the 'Shunt' Winding Flux then the windings are said to be **'Cumulative'**.

2) **Differential:** if the 'Series' Winding Flux **opposed** the 'Shunt' Winding Flux then the windings are said to be **'Differential'**.

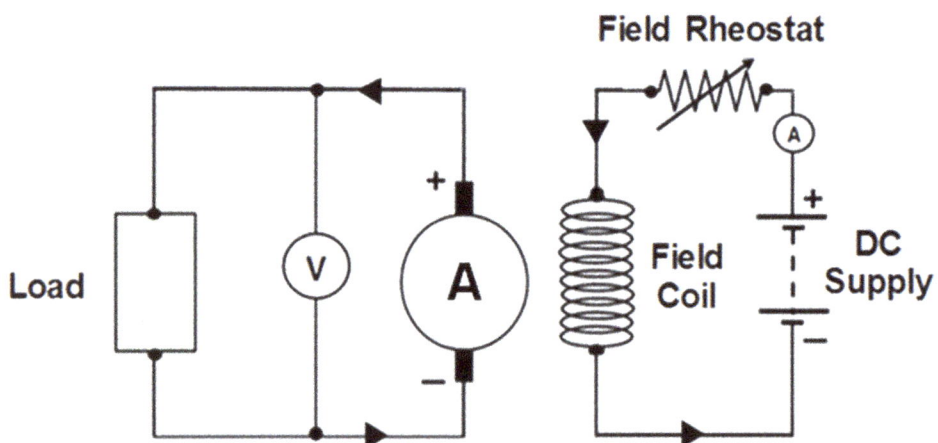

Fig. 22

The **Separately Excited** Alternators are where there is no actual Exciter Machine and is **NOT** Self-excited as a Generator and therefore must be excited from an external DC Power Source. The example in Fig. 22 uses a Battery Bank. An AC Supply through some sort of Rectification would also be suitable.

Not having an Exciter providing the Alternator Rotor with power, remember the spinning magnet in the case of this machine, can cause the flux density system to be a little more complex and in turn expensive. It must be mentioned that having the Exciter spinning on the same shaft as the Alternator and controlling the Flux Density via an AVR seems to be the norm rather than separately excited.

Permanent Magnet Generators (PMGs) as mentioned earlier are used to power an AVR and are in the design of a Dynamo where the Rotors are actually Permanent Magnets, and the Stator is a Coil. The PMG is not forced to be used to power the AVR in this case, the power can come from an external supply.

Voltage Increase can be achieved on the Alternator by being stepped up by a Transformer, as with Power Stations where this is done all the time. They generate a three phase at around 20,000+ volts AC and it is stepped up to 400,000+ volts AC by a Transformer for the Grid before it leaves the Power Station.

Constant Speed of the machine is maintained, and any adjustment is made to the Field Coil Flux Density. As the Generator accelerates in speed, the automatic Rheostat is adjusted to give the correct output as in Fig. 22. For ease, a Battery Bank has been used as the external power source, however other equipment may be employed which may make this method expensive.

Circuits as in Fig. 22 are commonly used on Electric Vehicle and Battery Charging Systems as Voltage Regulation and Output Current excellent The Ammeter is inserted into the circuit giving the Field Current which is easier to record than the actual Flux Density which would be very difficult.

1) **Unsaturated** Magnetic Poles the Flux is Proportional to the Field Current.

2) **Saturated** Magnetic Poles the Flux is Constant.

The Voltmeter is connected in parallel with the Generator and the Load and records the Generator Output Voltage to the Load as in Fig. 22.

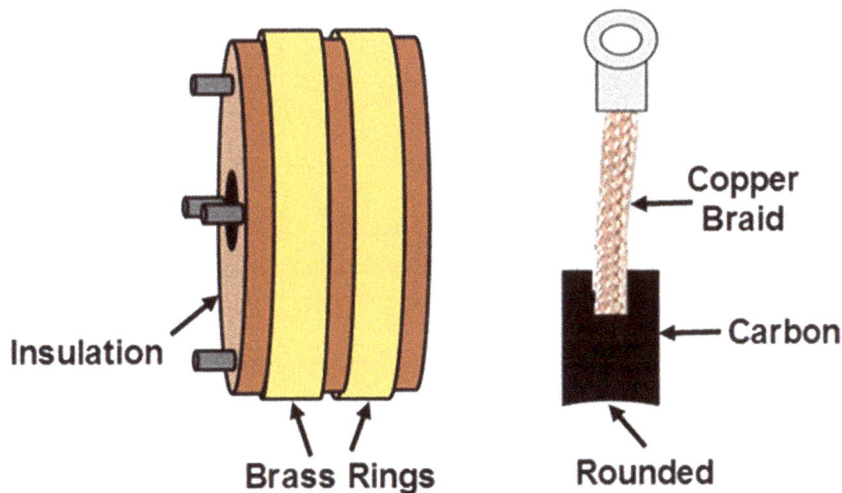

Fig. 23

Slip Rings together with Carbon Brushes as in Fig 23, is a way of transferring power **ONTO** a moving shaft to power the rotor, in the case of an Alternator, into an Electro-magnet, or transporting power **OFF** a moving Shaft, from the Armature, for the output, in the case of a Generator.

The rings are insulated away from each other by a material with a high dielectric strength (high insulator) such as resin. In the case of this Generator or Alternator there will, usually, be only two Slip Rings, but depending upon the duty and design, slip rings can have 2 – 8 Slip Rings in a line.

Slip Ring types that are not as common in use include Pancake, Fibre Optics, Wireless, Through Hole and Ethernet Connections. There were in the past slip rings called 'Mercury Contact Slip Rings' or 'Mercury Wetted' and these used liquid metal i.e. Mercury. The problem was that Mercury is not environmentally friendly, as well as being very toxic so these have been phased out. They may have other names such as: Electrical Rotary Joint, Swivel or Rotating Connector.

The Slip Ring Metal is usually Brass (an alloy of Copper & Zinc) and as in Fig. 23, there will be three Slip Rings if the Generator is Three Phase. Other metals that manufacturers can use include Stainless Steel, Steel, Cupro-Nickel, Bronze, Copper, Silver and even Gold, but are not likely to be used on those machines covered in this book.

Slip Ring Repairs are acceptable: for instance, if the surface has been slightly damaged, and if the repair is approved by the manufacturer. Turning the slip rings down on a lathe for instance to make them smooth and shiny, however this can introduce a whole set of new problems with brush pressures etc. Damage sometimes occurs where the rings become uneven all the way round. This may not be able to be repaired as the rings may be coming to the end of their life and a replacement set may have to be investigated.

Slip Ring Polishing can be undertaken, so long as the substance used is insulating and non-abrasive. Manufacturers will usually recommend the polish if one can be used on their brand of rings. To allow Rings to run at their most efficient, Slip Rings Maximum Speed must not be exceeded. This will be stated by individual manufacturers. Obviously when a machine is obtained from the manufacturers the Slip Rings that are on the machine should be correct and passed by their quality system.

Slip Ring Maintenance must be carried out regularly. I have carried out hundreds of Slip Ring and Brush inspections in the 1970s & 1980s when I was a young Electrical Technician. Checks may include:

a) How often do manufacturers recommend inspections?

b) Check the surface of the rings for discolouration or damage.

c) Check Brushes for wear.

d) Check Brush Pressure on the Rings.

e) Check the chamber for excess Carbon. (Remember Carbon is conductive)

Sometimes a thin film collects on the surface of the Slip Rings this is called **'Patina'** and it is quite harmless to the rings themselves. It is caused by the Carbon Brush riding on the Brass or Copper leaving a Carbon deposit mixed with Oxide compounds, dust & moisture in the surrounding atmosphere.

Brushes as shown in Fig. 23 are, together with Slip Rings, the way of getting power to or from moving Commutator or Slip Rings on a spinning shaft onto a stationary circuit. These Brushes must be made of a material that may wear itself but does not wear the Copper in the case of a Commutator and the Brass in the case of Slip-Rings. There may be more than one set of Brushes on the equipment.

Brush Material is usually Carbon (Element Symbol 'C') or Graphite, a type of Carbon, (Also Element Symbol 'C') and have a finite life. There are several different grades of material, so ensure that the correct grade is chosen as per manufacturer's instructions. Carbon and Graphite are very good conductors of electricity and at the same time stand the stresses, contact friction and heat whilst accomplishing their task. Brushes can also be made of Phosphor Bronze, but this must be done by manufacturer's design rather than customer choice.

Brush Holders are components which house the Brushes and ensure smooth movement onto the Slip Rings or Commutator without sticking as the Brush wears down. Brush Spring Pressure Mechanisms are on this Brush Holder to keep the Brush at the correct pressure, and they must be inspected at intervals not only to ensure the Brush length, but to also check that the Brushes are not sticking.

Too little Brush pressure could result in 'Bounce', causing excess sparking as the brushes bounce off the Slip Ring or Commutator. Too much pressure will result in unnecessary, excessive wear. It is possible for the Brush Pressure to decrease as the brush wears down, so it is vital that regular checks are done to ensure the correct pressure.

Brush Braid is what connects the actual Carbon to the stationary circuit as in Fig. 23. This Braid is embedded into the top of the Carbon or Graphite. If the Brush was allowed to wear down to this Braid and the Copper was allowed to ride on the Commutator, it is possible for it to wear a groove into the Commutator Segments as Copper rides on Copper, so Commutators should be inspected for grooves. If the Brush Braid rides on Brass Slip Rings a Copper trace will be left on the surface.

Brush Wear will speed up significantly if they are fitted to damaged Commutators or Slip Rings, because if they have rough surfaces through sparking, they will wear the Brushes more quickly. Brush Design must be changed like for like. Be careful with any other design or a brush that has been modified to fit as this could cause Brushes to stick or wear faster.

Brush Cleaning can be carried out, but nothing abrasive must be used and be careful using flammable lubricants as a fire could be started by the small sparks at the Brush tip when they are on the spinning shaft. Lubricant for Carbon Brushes should only be done with a manufacturer's approval. Carbon is self-lubricating to a certain extent, so should not require any additional lubricants.

Brush Chambers require cleaning out regularly as they get contaminated with worn Carbon or Graphite. Be sure to wear the correct PPE, i.e. Mask and Gloves, do not breath-in the Carbon. *Note: Carbon/Graphite has a very sweet taste and can be harmful if inhaled!* Sometimes, if the Brush Chamber gets impregnated with normal dust or grit, depending how abrasive it is, this can cause abnormal wear on the Brushes as well as damaging the Commutator or Slip Rings.

Brush Shape should be with ends rounded to the shape of Slip Rings or Commutator as per Fig. 23 to fit correctly. Flat ended brushes which are not rounded may increase 'Bounce or Chatter' and thus increase sparking. When bedding in the Brushes it is not difficult to make them the same curve as the commutator or slip rings. If the Brushes are not obtained already curved carry out the following:

a) If the Brushes are large, it may well be a good idea to file an initial curve into the Brush tip. Remember to cater for the people working around you with the Carbon Dust and on completion clean away the dust safely.

b) **WEAR CORRECT PPE AND DO NOT BREATH IN THE CARBON.**

c) Obtain a strip of smooth emery paper and put it round the commutator or slip rings. **Remember Coarse Side Up.**

d) Rest the brushes on the surface with the spring tension on the Brush Gear.

e) From below pull the emery strip backwards and forwards until the correct curve is worn into the Brush Tip.

Vibration can affect Brushes and reduce their life. Very severe vibration can increase 'Bouncing' of the Brushes on the Slip Rings or Commutator. This may also have a detrimental effect on the Slip Rings, Commutator, Brush Holders and Spring Mechanism.

Excessive sparking on the Commutator can be caused by the Brushes not being positioned in the **'Neutral Plane'.** They must be in contact with the correct segments of the Commutator, otherwise they can short out other segments and short circuit the windings, which will cause a lot of sparking at the brush tip and possibly cause heat and winding problems.

Lots of sparking at the Commutator or Slip Rings will cause them to appear dull and rough to the touch and will also wear down new brushes fairly fast.

Commutator Wear can also happen where the Copper wears down on the Commutator itself, and in this situation, it is not the brush wear that is the problem. If the Copper wears down too much on a Commutator and the Mica, which is an individual segment, insulation becomes exposed, and this can have a detrimental effect on the Brushes making them bounce over the Mica protrusions.

Look for very fine deposits of Copper on the Brushes which might be an indication that the Copper is wearing! This fault is not an easy fix and it may actually require removal of the Armature for repair or at the worst renewal.

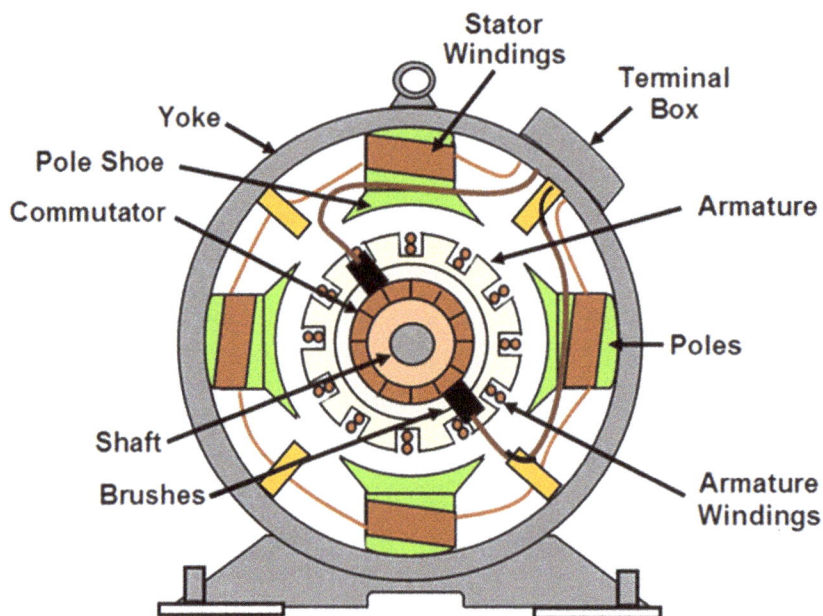

Fig. 24

Insulated Laminations are what solid components such as, Field Poles, Armatures and Rotors (shown in Fig. 24) are divided into, to stop Eddy Currents forming in them when they are running in a Magnetic Field.

Eddy Currents are Loops of EMF that flow when any magnetic circuit is working. How the manufacturers stop them is to chop the Solid Metal Components into insulated laminations which are thin strips of metal.

Armature Reaction is a phenomenon that can occur in Magnetic Fields. Without going too deep into a complex subject: An Alternator (Synchronous Generator) has two separate windings:

a) The **'Field'** winding provided by the Main Stator Field Coils.

b) The **'Armature'** winding which is the Rotor.

Armature Flux is formed when the armature provides a separate flux of its own, due to current flowing in the coils, to the main field winding magnetic flux.

It is possible for this Flux to become distorted, and if not corrected would have a detrimental effect on the Main Flux and this is '**Armature Reaction'** and may also have an effect on the Alternator Efficiency.

On a DC Generator this Armature Reaction can have another effect in that it can move the **Neutral Plane.**

a) The Power Factor will have an influence on the Armature Reaction and vice versa.

b) Movement of the Brush position in relation to the Neutral Plane would also affect the Armature Reaction.

Generator or Alternator Assembly is the next learning stage after the basics of how they work. Where does the information that has been discussed up to now fit into an actual machine? In the next section we will look at the build-up of the machine.

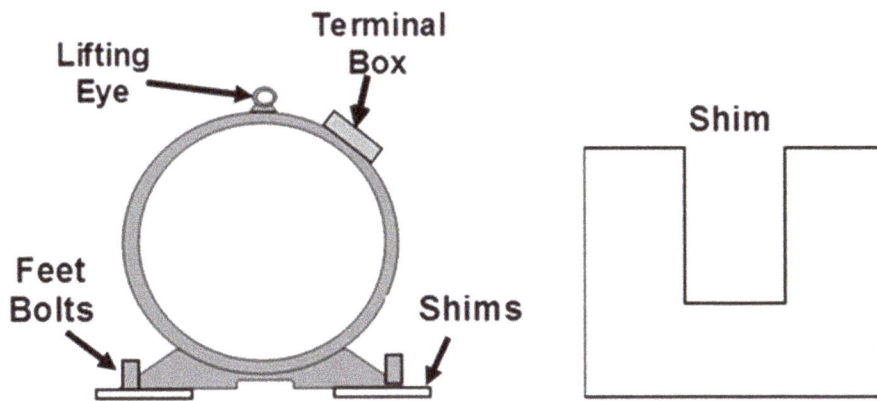

Fig. 25

The Yoke is the frame of the Generator or Alternator and if the machine is around the size of a large Electric Motor, then it will be a similar design as Fig. 25 above it is made of cast iron and holds the Poles, Field Coils, Rotor/Armature, Terminal Box, Bearing Housing and Brush Gear. Power Station Generators and Alternators are much larger and will be fabricated in many metals.

Yoke alignment is vital between this and the Driver, let us say of a Steam Turbine, at the coupling, otherwise misalignment could result in severe vibration. Shims, as in Fig. 25, are fitted under the actual feet of the machine, be that the Generator, Alternator or Steam Turbine. The Shims, which are very thin sheets of metal, are there to accurately lift the Yoke or the Driver to be exactly level on the same plane.

In the past, alignment would have been done by an Electrical or Mechanical Technician with a straight edge, but in modern times they will be Laser Aligned. When a small Generator with Shims is removed for any reason, it may be necessary to fit new ones, but old ones should be kept on the bed, wired to the foot bolts to give an idea of how much the machine had to be raised before removal.

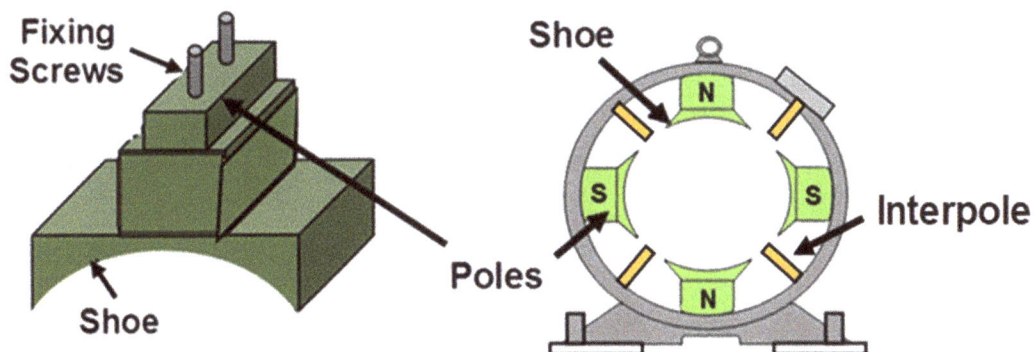

Fig. 26

Let us now talk about the Field Poles as in Fig. 26. The Yoke, you will notice, has two pairs of Poles which house the Stationary Field coils, which in a Generator provides the Magnetic Flux (Magnet) and in an Alternator is the Stationary Armature (Coil). Pole Cores are another name for these blocks and, as mentioned earlier, in any Magnetic Flux situation the actual poles are not solid, they are laminated to stop **'Eddy Currents'** from flowing through the solid metal and causing heat.

Inter-poles are positioned between the Poles as in Fig. 26 and are connected in series with the Armature helping in reducing Armature Reaction. They would have the same polarity as the Pole ahead of it and would correct the position of the Neutral Plane should it be moved by Armature Reaction in the Generator.

Pole Shoes is the name given to the end of the Pole Core, which is curved to accept the round Rotor/Armature. It is essential for manufacturers to get the Air Gap correct between the Armature and the Pole Shoes. When spinning, the Armature must not touch the shoe but must be very close. In an Electric Motor the Air Gap will lower the Torque if it is too large. In a Generator or Alternator, efficient interaction between the Magnetic Flux Density, Reluctance etc. from the Armature and the Main Field relies on a small Air Gap.

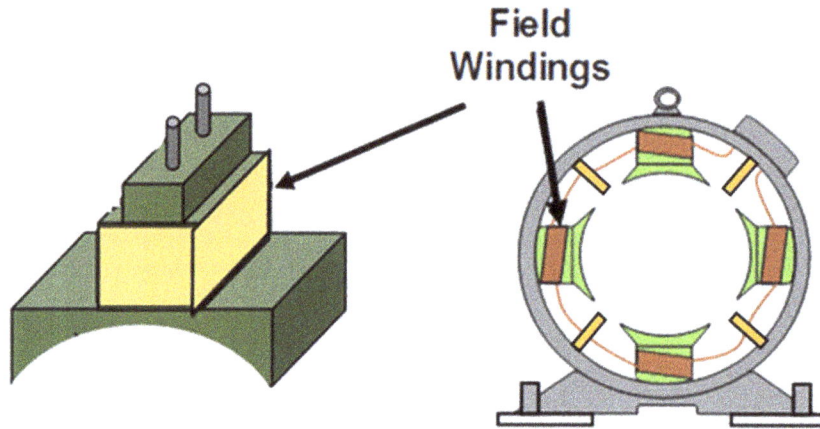

Fig. 27

Stator Field Coils are wound onto the Pole Cores. These are the Copper Wire Coils as mentioned in Fig. 27 above. These form a Field Magnet in a Generator and a Stationary Armature in an Alternator. The Pole Cores holding Field Windings are then installed into the Yoke and terminated in the Terminal Box on the upper right. The Pole Core provides a low Reluctance pathway for any Magnetic Flux.

Fig. 28

Do they call it a Rotor or an Armature? From the Information up to now Fig. 28 above, is an **'Armature'** with the Coil Wires that are fitted into the Yoke of a **'Generator'** and in this case the Generator output would be DC so the Armature would have a Commutator fitted to obtain the DC Output. A **'Rotor'** would be a piece of equipment that is fitted into an Alternator. This Rotor, as mentioned earlier, is the device in an Alternator that becomes a Magnet spinning in a coil with its power coming from another DC Generator called an Exciter.

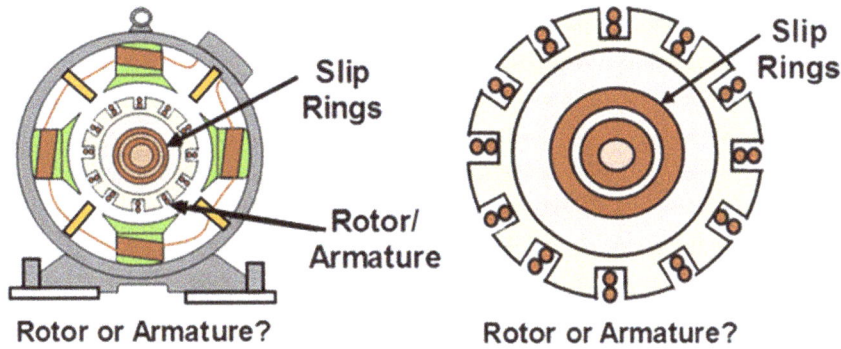

Fig. 29

The above Fig. 29 shows an **Armature** where it is fitted into the Yoke of a **'Generator'**. The Generator output would be AC via the Slip Rings with no Commutator to make the output DC. So here there would be an Armature Coil spinning with Field Coils forming the Magnet. I have drawn the Slip Rings one larger than the other to remind you that there are two.

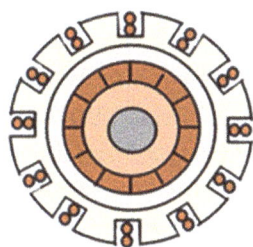

Cylindrical Rotor

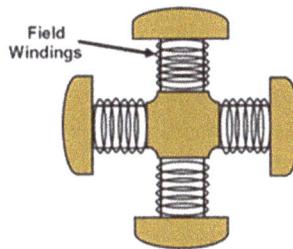

Fig. 30

Salient Pole Rotor

Fig. 31

Rotor Types depend on the Machine, and it all depends upon the duty i.e. A high-speed Cylindrical Shape. (Fig.30) and a low speed **'Salient Pole'**. (Fig. 31) Where it is used in an Alternator, its shape would be very thin with a very large diameter and of course accompanied by its Exciter.

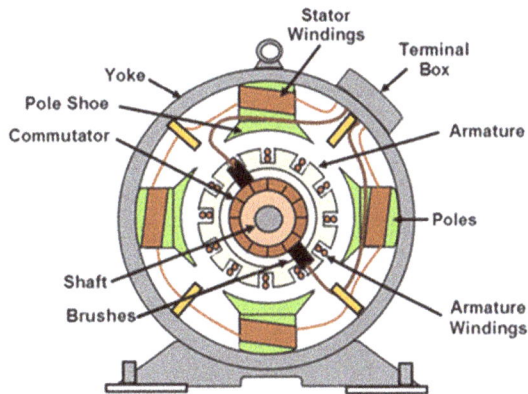

Fig. 32

Permanent Magnets could be used for the **'Field Poles'** of the above Generator and in Alternators Permanent Magnets could be used for the **'Rotor'**. There would be no worry in this case about Residual Magnetism as the Magnetic Flux would always be present, but Permanent Magnets as they get older will lose some of their Magnetic Flux Density and efficiency will go down as the field weakened. There is no way we can counteract this weakening, we cannot re-magnetise without fitting new magnets, hence the use of electromagnets instead.

Residual Magnetism as we discussed earlier under Winding Configurations Fig. 17, can be very confusing. It is possible for a Generator to run but not produce electricity when there appears to be nothing wrong with it. The probable reason is due to lost **'Residual Magnetism'** sometimes called **'Remanence'**.

Residual Magnetism depends on how much **'Retentivity'** a material has, to retain a certain amount of Magnetic Flux. Lingering Magnetic Flux must be present from when the Generator last ran to enable it to produce electricity on the present run.

If the Magnet involved in the magnetic field was an actual Magnet and not an Electromagnet, then Residual Magnetism would not be a problem, but where magnets are used, it is called a 'Permanent Magnet Alternator.' However, if the machine is an Alternator fed by an Exciter to make its Rotor into a Magnet, there would not be this problem.

Residual Magnetism can be lost if:

1) The Generator is shut down for a good length of time.

2) The Generator is run for too long without any load.

3) The Generator was shutdown with a large load still connected.

ALWAYS CHECK WITH THE MANUFACTURER FOR THEIR METHOD OF RESTORING RESIDUAL MAGNETISM!

The Service Factor is usually stated by the manufacturer. This is the maximum load that the Generator can handle over 100% of its full load for a short time. Too much time over 100% load will damage the machine. Increases in Generator load past 100% will raise the temperature and current, which of course could cause the windings to short circuit and burn out.

There will also be a Service Factor for the driving machine driving the Generator, and again the machine manufacturer will supply the figures. The two questions here must be:

1) Why is the system overloading beyond 100%?

2) Is the Generator too small?

Domestic Generators are thought by many to do more than they will. The common standard Domestic Inverter-Generator provides around 2KW (2000Watts). To run the average house, a Generator capable of providing 7KW – 9KW (7000Watts – 9000Watts) would be required. Domestic Generators would usually be Single Phase.

Generators or Alternators in Power Stations are huge and generate three phases of power at 20,000+ volts, but regardless of size the basic principles that have been discussed up to now still apply. Using Brushes and Slip Rings for large loads can be done, but large loads would be better using Alternators where the Rotor is a Magnet spinning in a Coil, and the load is taken from the Stator and not from the Rotor.

To sum up two very important points:

1) Generators are a coil spinning in a magnet and require Residual Magnetism from when they last ran to enable them work.

2) Alternators require an Exciter, Automatic Voltage Regulator (AVR) and Permanent Magnet Generator (PMG) to enable them to work. Called Synchronous Generators.

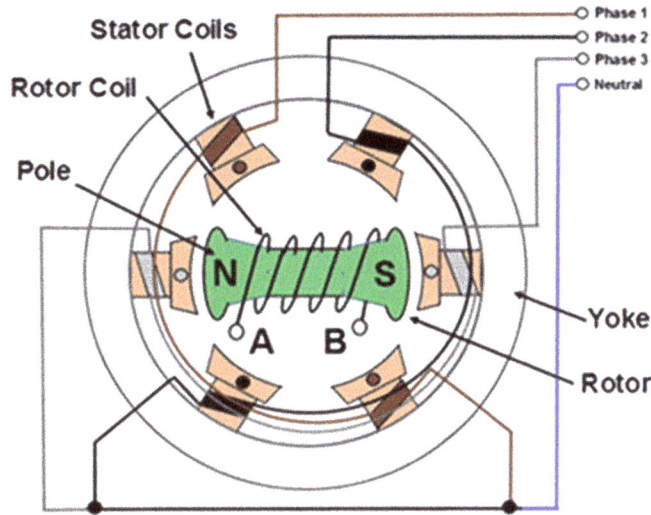

Fig. 33

Three Phase Field Coils would have windings 120° apart, the total of all three making 360°, as in Fig. 33 above. This would have an electromagnet fed by slip rings **A & B** from the Exciter Commutator spinning within the Field Coils.

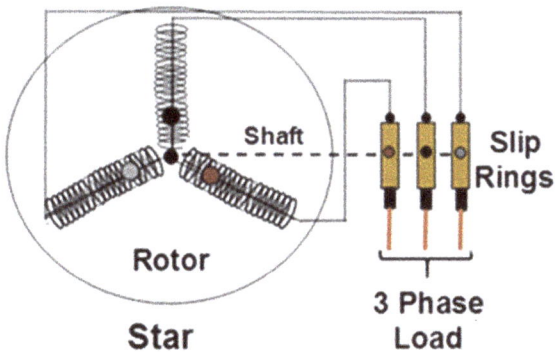

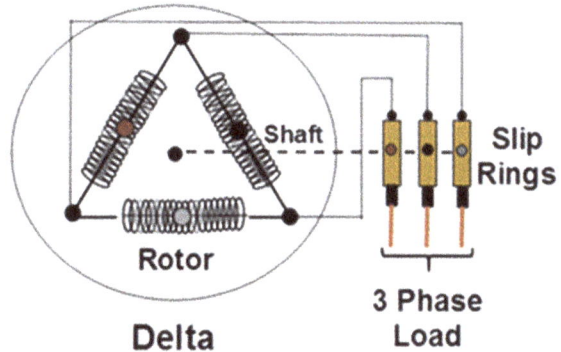

Fig. 34   **Fig. 35**

Star or Delta Stator Coils are installed in 'pairs' of coils. Several coils are connected for each phase with the other end of the coil connected in a **'Star'** as Fig 34. Star is the connection that most Generator Rotors with Slip Rings are **ALWAYS** preferred to be connected into as per Fig. 34. Star conveniently provides a Neutral.

Delta Neutrals can be provided with an **'extra'** Star Winding would have to be added. One more promising item about the Star connection is that the Phase Voltage is less than the Line Voltage so armature windings will be less.

Generator Standby Systems (GPS) are special start-up systems that automatically start up the Generator(s) in the event of a power cut. The system changeover is done by what is called a **'Static Switch'** which is sometimes called an **'Automatic Transfer Switch'** (ATS) and its job is to transfer from Standard Supply to the Standby Generator Power Supply (GPS) in the case of a power outage. These switches are instantaneous.

The GPS System may be used in hospitals where emergency standby power is essential for example in an operating theatre. This GPS must be assessed against an Uninterruptable Power Supply (UPS) to see which system is the most appropriate.

Fig. 36

Synchronisation of an Alternator or Generator to a Grid System must have certain conditions met. These days this is most likely done using an Electrodynamic Synchroscope as in Fig. 36. This instrument has one section monitoring conditions on the Grid and a section monitoring conditions on the machine and compares the readings to see if they match.

Connecting two supplies together with different parameters that do not match can severely damage equipment due to a high current, if it is not protected. The Voltage and Frequency must be the same on the Grid as it is on the Generator to be connected, otherwise Reaction Forces may damage the machine.

The Synchroscope (Fig. 36) needle rotates anticlockwise if the machine rotation is 'Slow' and when the needle rotated clockwise the machine rotation is 'Fast'. When the needle is stationary and pointing upwards the Speed and Frequency of the machine is matched to the Grid.

The Phase Sequence of the machine output must match that of the Grid it is going to connect to. Phase Sequence is the order in which the phases are placed 120º apart, i.e. Brown, Black & Grey on the Grid must match Brown, Black and Grey on the Generator. So, for instance the Grid could not be Brown, Black & Grey and the Generator be Grey, Black & Brown otherwise there will be a tremendous, short circuit when the circuit breaker is closed.

The Phase Angle of the machine must match Phase Angle of the grid. The Phase Angle here is the cosine of the angle between Voltage and Current and should be kept as near to one (Unity) as possible. The Phase Angle is also the Angle between **'Real Power'** and **'Apparent Power'** as we discussed earlier under 'AC current'.

Generator/Alternator Faults Sum up:

There are many common faults that can occur with an Alternator. One of the main devices that would certainly give problems is a faulty AVR. This could manifest itself in several ways, for example, the Alternator output may be high or low due to the AVR not correcting the Exciter causing excitation to be lost or reduced. So, for example, the Alternator could shut down on under voltage.

There are several ways that the AVR could malfunction. An example could be incorrect information from the Voltage or Frequency Sensors, which might cause the Alternator Speed to fluctuate or over-speed.

Should the Circuit Breaker to the load trip, this could also cause the Alternator to over-speed. The AVR might have lost its supply link to the PMG, or the PMG might be faulty. The Alternator could be running with no output. Armature Reaction could also increase. Another AVR problem, also linked with the above, could cause the Alternator to 'Hunt' or 'Surge', trying to find a new stable equilibrium, but this can also be linked in with the Load 'Surging' or 'Fluctuating' badly.

As we have already discussed, Brushes can cause excessive sparking if, for example, the Brush is not in the Neutral Plane on the Commutator, meaning it is not in contact with the correct segment, or Brushes not rounded to ride correctly on the Commutator or Slip Rings.

Incorrect Brush Pressure is another problem. This is where the pressure is too light and the Brush is subject to 'Bounce', especially if the machine is under a lot of vibration. Other problems which can cause excessive sparking at the Brush Tip are, Brushes sticking in their holders, and something catastrophic happening to the Armature Windings.

On an inspection if it is found that the Brushes are wearing extremely fast, this can be the result of several problems. The Brush Spring Pressure could be too great and pressing the Brush down too hard; there is damage to the Commutator or Slip Rings causing the Brush to wear or, the incorrect material Brush has been fitted, and the Carbon is too soft.

Commutator or Slip Rings. Damage to the Commutator or Slip Rings can happen for several reasons. They may have come to the end of their life and need to be replaced. The Brush may have worn down too far and exposed the Copper Braid allowing it to ride on the Commutator. This would in actual fact wear a groove in the Copper of the Commutator which may show as copper filings on the Brush. Brush Bounce will also cause Commutators or Slip Ring damage by sparking.

Cannot Synchronise the machine to the Grid. Firstly, check several parameters which need to be matched from the Machine to the Grid, these being Voltage, Frequency, Phase Sequence & Phase Angle. The Synchroscope should supply the answer to some of these figures. Safeguards should be built into the system to ensure that the Circuit Breaker or Static Switch cannot close if any of the above parameters are not matched.

Generator running but no Power Output. After checking equipment i.e., has the Circuit Breaker or Static Switch closed? If the Machine is a Generator, has it lost its Residual Magnetism and if so you will require a car battery to re-magnetise, as per manufacturer's instructions.

If the Machine is suffering from severe vibration, it can be very concerning as to the actual cause. If the Machine has run some time without a problem, the likelihood of the Driver and/or the Generator being out of line is not high. A potential catastrophic bearing failure is another problem that can be checked with instruments. Another check might be to see if the Windings have become unbalanced or lost a phase?

If the Machine is running hot, check how hot against what would be the norm. There could be a problem with the Cooling System, so check for obstructions etc. If the Machine overloading is due to an increase in load etc. Check to see why have the overloading systems and instruments have not operated?

We always seem to return to, is there an AVR fault or something like a sensor not giving the AVR the correct information. However, heat can also be caused by something like a winding fault inside of the machine. Bearing collapse is another problem causing the Rotor to ride on the Field Poles.

Sensors on the Alternator, such as Voltage and Frequency etc. will feed incorrect information causing the AVR problems in controlling the Exciter Field. Being such important devices which although would not be very costly in the circumstances, it is worth having some spare tested sensors ready to be installed.

Important Definitions:

1) **Rotor:** Spinning Rotor Magnet of an Alternator fed DC from the Exciter.

2) **Field Coils:** The outer Stator Coil of an Alternator. Stationary Armature.

3) **Armature:** Spinning Coil of a Generator.

4) **Field Winding:** The outer Stator Coil of a Generator forming the Magnetic Circuit.

5) **Exciter:** A DC Generator that feeds DC via Slip Rings onto the Alternator Rotor.

6) **Spinning Diodes:** A non-sparking alternative to a sparking Commutator.

7) **AVR:** Controls the Exciter Field, Flux Density and in turn Output in the Alternator.

8) **PMG:** Supplies the AVR with power.

9) **Synchronous Generator:** Another name for an Alternator.

10) **Asynchronous Generator:** Machine does not operate at Synchronous Speed.

11) **Slip Rings:** Devices to obtain power to/from a spinning shaft to a stationary circuit.

12) **Residual Magnetism:** Required by a Generator to start generating.

Fleming's Rule:

John Ambrose Fleming (1849-1945) invented a Rule, where inside of a Generator, certain information can be obtained using a very simple action which involves your right-hand finger and thumb formation as per Fig. 37 below. Remember here, that we are on a Generator not a motor.

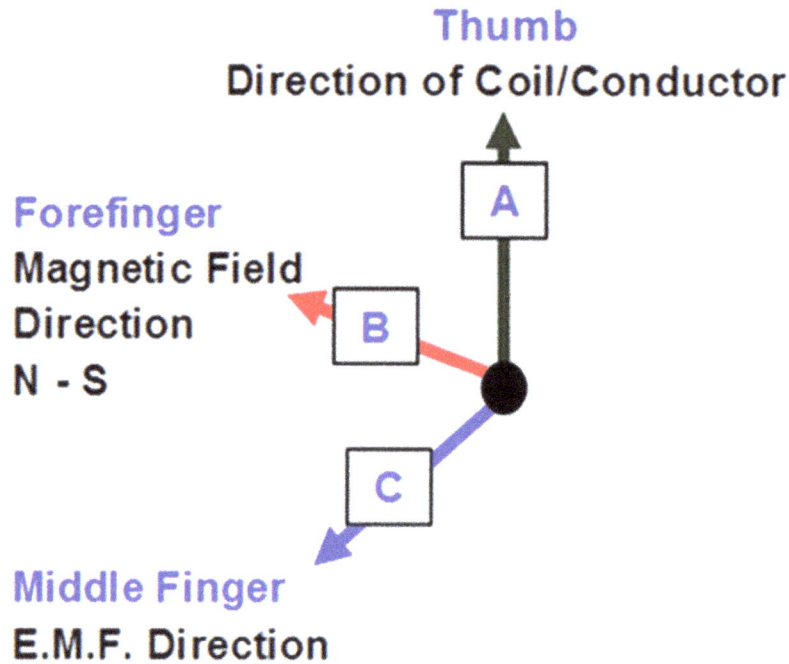

Thumb
Direction of Coil/Conductor

A

Forefinger
Magnetic Field
Direction
N - S

B

C

Middle Finger
E.M.F. Direction
Fig. 37

Fleming's Right-Hand Rule reads: Firstly, place your Thumb, Forefinger and Middle Finger as per the diagram in Fig. 37 above.

A: Being the movement of the Armature.

B: Being the Magnetic Field Direction North to South.

C: Being the EMF Direction.

Remember that this is a driven machine **NOT A MOTOR!** If this were a Motor, we would be using Fleming's Left Hand Rule. Try the exercise on the next page.

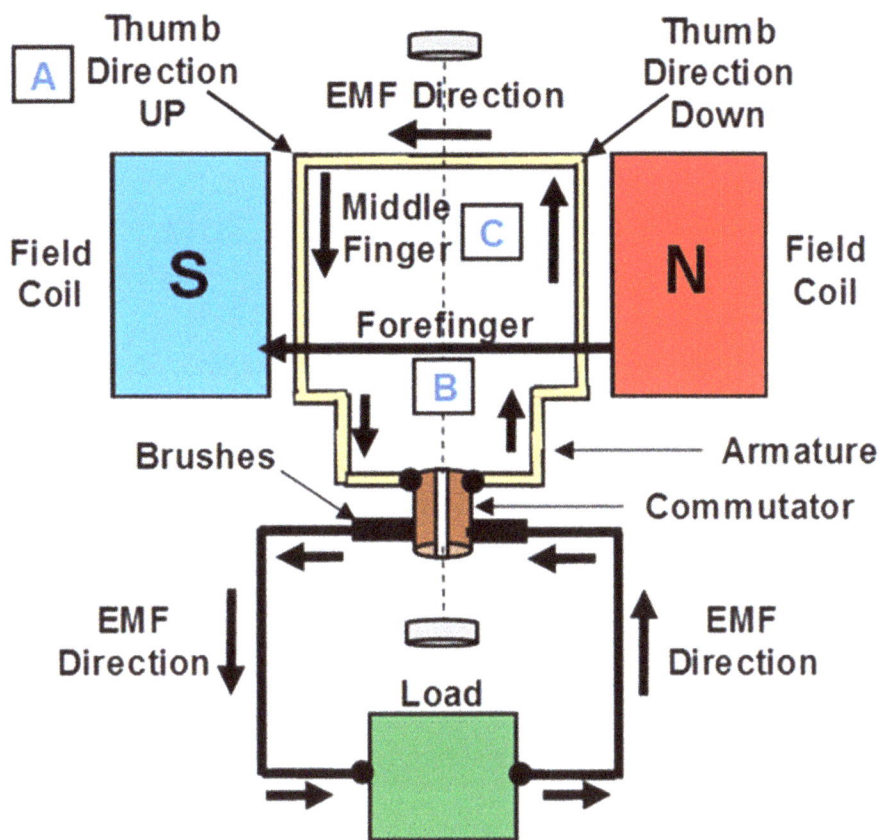

Fig. 38

Using Fleming's Rule, look at diagram Fig. 38 with your fingers as per Fig. 37 and follow the steps below:

1) The forefinger **(B)** points towards the direction of the **Magnetic Field** (North to South) these are coloured Red for North and Blue for South in Fig. 38.

2) The middle finger **(C)** points towards the **direction of the E.M.F.** As you can see this is Anticlockwise in Fig. 38.

3) Thumb (A) points to the movement direction of the coil. With your Forefinger pointing from North to South and your Middle finger in the direction of the EMF which is Anticlockwise.

4) Looking at your thumb you should find that the left-hand side of the Armature will come **UP** towards you out of the page. The right-hand side of the armature will go **DOWN** away from you into the page.

Fleming's Right Hand Rule applies to Generators. To help to remember which rule, we call them **GENERIGHTERS**.

a) Fleming's Rule applies to most Generators with Armatures and Field Coils.

b) His Left Hand Rule, of course, applies to Motors.

Now try experimenting and see that it works with other parameters. In your mind Change the direction of the EMF and now use Fleming's Right Hand Rule and see what you find in direction of rotation.

Now change the poles of the magnet in your mind or draw it out and apply Fleming's Right Hand Rule and see what you find.

Induction Generator:

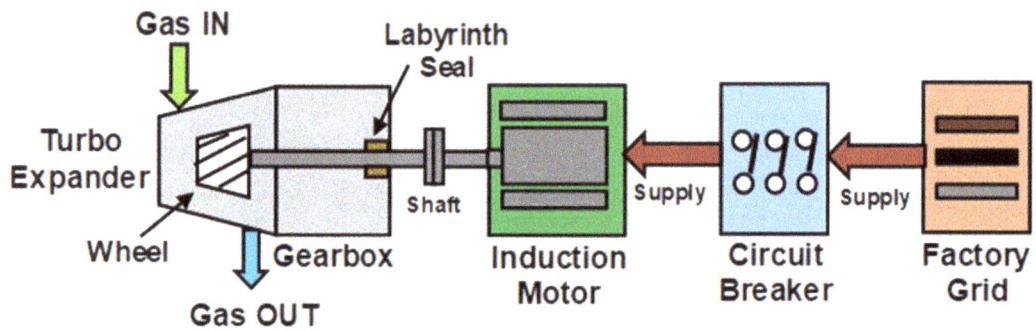

Fig. 39

The Induction Generator is basically a Three Phase Induction Motor, Fig. 39, which has a Drive Machine connected to drive the Rotor past the Rotating Synchronous Speed on the Stator, at which point it generates an induced EMF into Stator Windings, and it turns into a Generator. This is one of the simplest Generators.

Another name for this Induction Generator is an **'Asynchronous'** Generator, as opposed a **'Synchronous'** Generator, because it operates at a different speed and waveform to the Synchronous Speed of the Machine. The Induction Motor Rotor can be a Three Phase Squirrel Cage Induction Motor or, it can have a 'Wound Rotor' with windings instead of Rotor Bars. These are more efficient, but require Slip Rings and Brushes, see Fig. 40.

The Induction Generator relies on the fact that it starts as an Induction Motor so is already connected to the Grid. The Drive Machine, which in this example is a Turbo-Expander, now drives the Rotor faster and passes the 4% **'Positive Slip Speed'**. It then catches up to rotating **'Synchronous Speed'** and when it reaches this point, no torque is generated from the Motor. As it passes Synchronous speed it turns into a Generator, and this would be called **'Negative Slip Speed'** and here the Rotor Flux is cutting the Field Coils.

Reactive Power is required by an Induction Generator and is achieved by being connected to the grid; this is called **'Grid Excited'**. Stand Alone Induction Generators are not usually an option, as they would have to be connected to a Capacitor Bank to provide the **'Reactive Power'**. This is called self-excited and a loss of the Grid connection whilst running will cause a loss of Generation on the connected Generator.

The Rotors of Squirrel Cage Induction Generators do not particularly like the speed of the driving machine to fluctuate, otherwise extra equipment may be required such as an Inverter/Converter System. This phenomenon is not so bad on the Wound Rotor machines.

Over-speed is not usually a problem when the machine is connected to the grid, but there must be over-speed systems in operation just in case the Circuit Breaker was to trip and, with no Reactive Power Load, the Induction Generator may over-speed with the Turbine still driving.

Can a Standard Induction Motor be used as an Induction Generator? Well, the answer is **YES** and **NO**, the smaller motors will produce a small generation, but nothing that can be used. Motoring is a term used when the Fan or Pump that is being driven is passing slightly and spinning the Motor. It will be found that an IR or Balance Test cannot be done on the Motor because it is generating a small voltage thus slightly affecting your IR/Ohmmeter, in fact damage can be done to the instrument if its buttons are pressed.

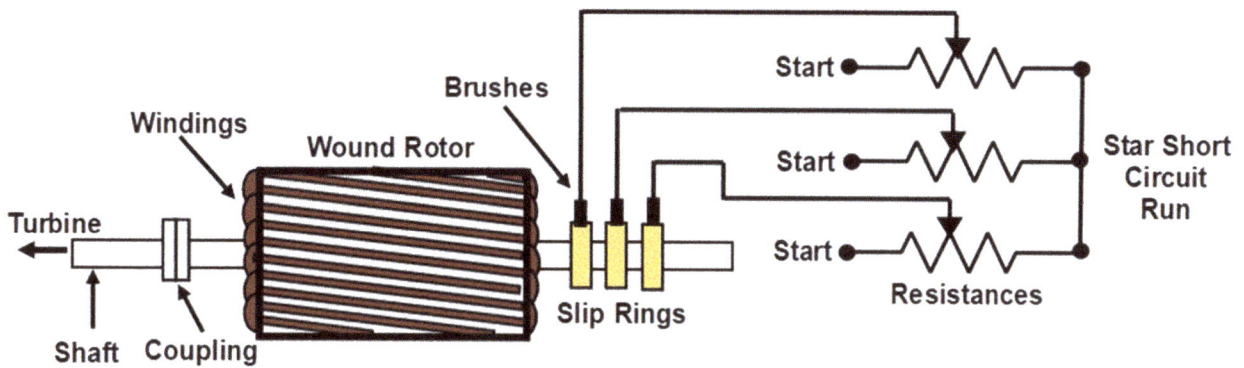

Fig. 40

The Driving Machine can be a Gas Turbine, Turbo Expander, Hydro Power, or a Wind Turbine Engine etc. Fig. 39 shows the example of a Turbo Expander now driving the Induction Generator where the output is now onto the Factory Electricity Grid via the Circuit Breaker.

The Turbo Expander Speed can be around 30,000 RPM. If this is the case, then the speed would have to be dropped considerably to around 1500 RPM to drive the Generator and this would be achieved through a gearbox.

Induction Generator Synchronisation is not required if it is a 50Hz Machine feeding onto a 50Hz grid. Remember to start off, it will be an Induction Motor. The Rotor Bars have very little resistance and are short circuited at the ends of the Squirrel Cage, so a high current is the result here.

The Rotor Current here, is in fact, causing a magnetic field which is cutting the Stator Coils. Sometimes these Induction Generators when driven by a Turbo Expander, are there with the sole objective of putting a load on the Driving Machine and not just to generate huge amounts of electricity.

Squirrel Cage Induction Generator Output is taken from the Stator Terminal Box as there is nowhere else for it to come from. Remember that this really is an Induction Motor being used for another purpose. The 'Squirrel Cage' part of the motor's title gets its name from the Rotor without any 'bars' fitted, and t resembles the exercise wheel that is used for pet hamsters.

Soft Start Units can be used to 'Soft Start' the Induction Motor, and where necessary a Wound Rotor, as in Fig. 40., if there is a large load connected to the coupling, such as a Wind Turbine Rotor. For this Soft Start the Wound Rotor would be used with the Slip Rings, possibly going to a Rotor Resistance Unit. For the motor to attain full speed the rotor resistance would have to be wound fully out and the 'Star Short Circuit' connecting the windings in Star with no resistance. Resistances are not used during generation.

The word **'Induction'** in both titles, **Induction** Motor and **Induction** Generator, suggests that this may give a not so good Power Factor where the current is lagging the voltage. Quite a few of these running together would severely affect the Power Factor of the whole factory. The question arises "can we do anything about it?" The answer is Yes, we can, by adding Power Factor Correction in the form of Capacitors.

Induction Generators with the Squirrel Cage can be obtained which are Atex Approved and are quite suitable to operate in a Zoned Hazardous Area. The Wound Rotor ones with Slip Rings may not be suitable in a Hazardous Area.

Magneto-hydrodynamic Generators:

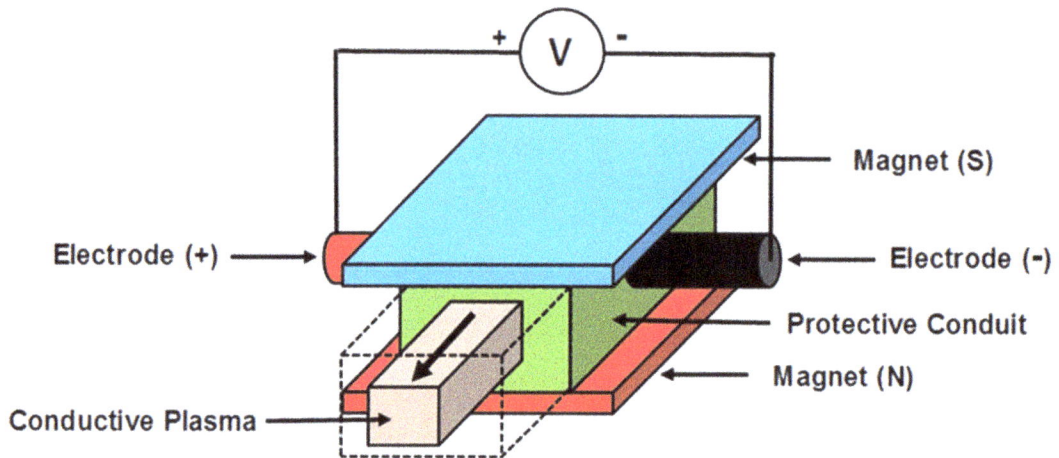

Fig. 41

The Magneto-Hydrodynamic Generator (MHD) was invented by Hannes Alfvén who was a Swedish Electrical Engineer & Plasma Physicist, born in 1908 and died in 1995. Also known by its other name of a **'Fluid Dynamo'** it is not an actual Mechanical Generator but is around 50-60% efficient.

Conductive Plasma flows very fast through the Magnetic Field produced by the two magnets, which are shown in the diagram as Red and Blue and there are literarily no moving parts, except of course the Plasma. Electricity is produced at the Black and Red electrodes in Fig. 41 and when the plasma flows through the protective, Green, conduit it is at a very high temperature of between 1500°C - 3000°C.

The electricity is Direct Current, DC, of course, but we can change it to Alternating Current, AC, by passing it through a device called an Inverter.

Plasma, which is an electrical conductor, appears to be a very mysterious substance and the most common state of matter in the universe. In fact Plasma is the 4th state of matter, after Liquid, Solid and Gas. If I can describe it at all, it is a sort of ionised soup. Its temperature is so hot that all the atomic bonds have broken down and it is full of charged particles and easily affected by magnetic fields.

There are several Engineers of the past, who had a hand at determining the laws of electricity production, starting with Lorentz Force Law which discusses the effect of a charged particle moving in a constant Magnetic Field.

Michael Faraday's Law of Electromagnetic Induction which states the movement of a fluid through a Magnetic Field Induces an Electric Current. Faraday did an experiment using charged particles in the water of the River Thames.

Fleming's Right Hand Rule still applies here because the movement is the Plasma, and there is a Magnetic Field from North to South and the EMF.

Is this Generator Environmentally Friendly? This is the big objective of modern day, and the answer is Yes, as far as the actual Generator is concerned, although if a high pressure and temperature coal gas is used to fuel the combustion, then this part of the process is questionable.

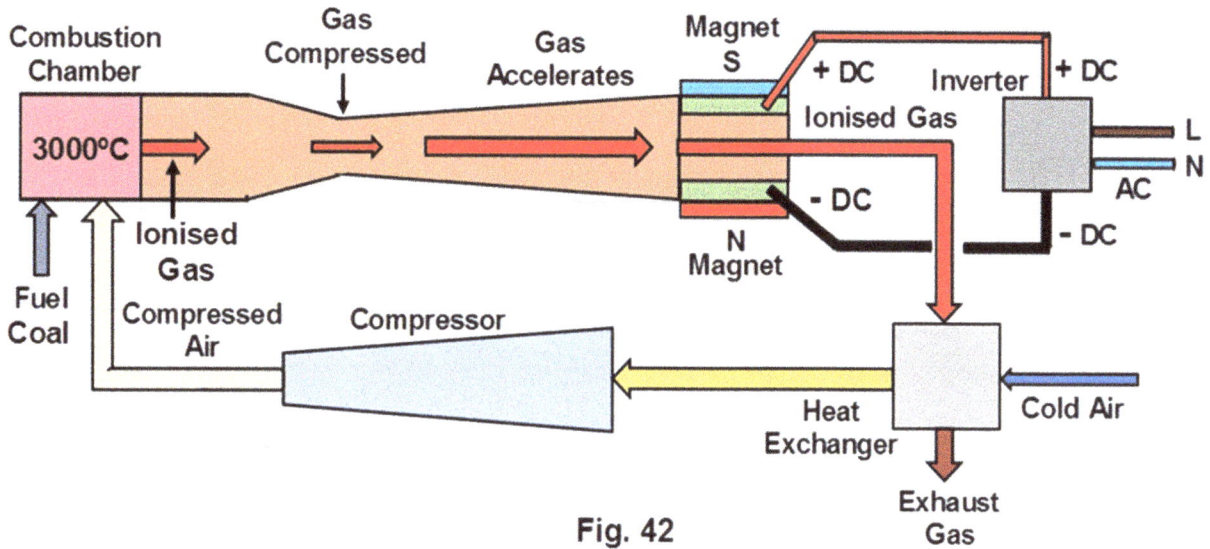

Fig. 42

The full process is as Fig. 42 above. In this case we are using Ionised Gas, which is a type of Plasma. The Gas is heated to a very high temperature in a Combustion Chamber causing atoms to collide violently and tearing away their electrons. It is then compressed in a Cooling Tower Shaped Chamber (see Fig. 42) which makes it speed up to go through the generator section, at around 3000ft/sec, as in the top diagram.

There are currently three main types of MHD Generators:

1) The Faraday Generator.

2) The Disc Generator.

3) The Hall Generator.

There are two different methods of MHD:

1) The **Open** Cycle Method: What has been previously described.

2) The **Closed** Cycle Method: A Working fluid, which could be liquid metal, is circulated in a closed loop.

The **Open** Cycle Method is shown in Fig. 42, where fuel is burned in a combustion chamber at around 3000ºC and when mixed with pressurised air from a compressor, forms plasma which is then directed through the process.

The **Closed** Method Cycle uses a process of liquid metal or inert gas instead of plasma in a closed loop, which is constantly used in the system. The liquid metal used is **Potassium (Chemical Symbol 'K' for Kalium – Latin name for Potassium)** which is heated in a Combustion Chamber to obtain its heat and is then passed through the cooling tower shaped chamber to give it some velocity. The inert gas that can be used is **Helium (Chemical Symbol He).** Either of these would take the place of the Plasma.

In conclusion, what we have here are outer magnets which form the Magnetic Field similar to the Field Coils in a conventional Generator and instead of a spinning Armature moving within the field, we have moving Plasma. Sometimes called an MHD Converter, it has no moving parts. Can it be used in a Hazardous Zoned Area? Well, the problem would arise with the Combustion Chamber along with a Plasma that could reach 3000ºC. If the fuel was coal to obtain the heat, then, of course, it would be Non-Renewable.

'Free Piston' Linear Generators:

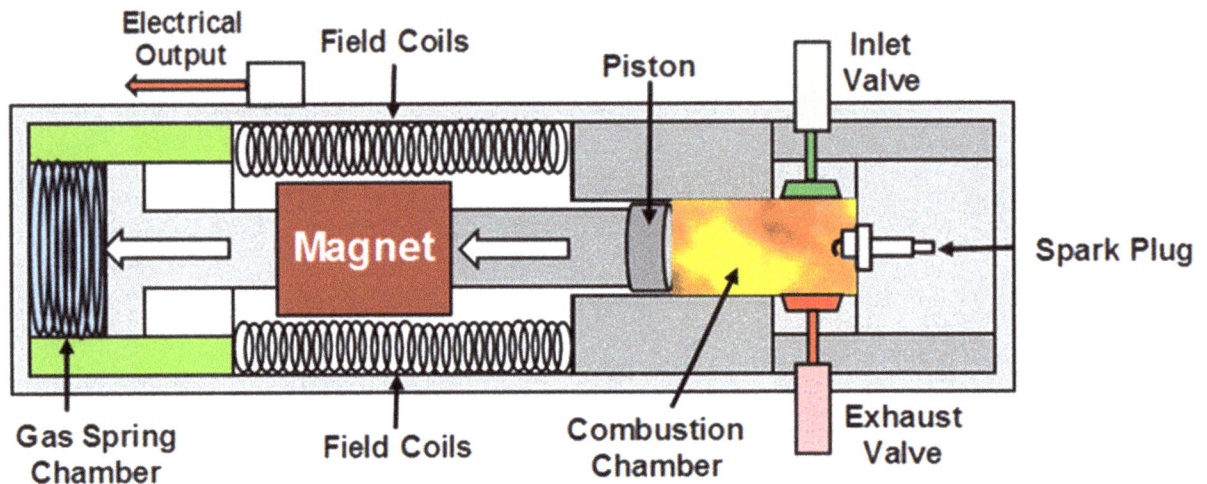

Fig. 43

The Combustion Engine is the basis for this Linear Generator, patented in 1940, shown in Fig. 43 and Fig. 44, which at present would run on Fossil Fuel, Petrol or Diesel, so Carbon Monoxide (CO) is the biproduct.

The good news is that they now have a design of Linear Generator to run on Hydrogen and Biofuel which is an environmental break through. However, it depends on which Biofuel they use, but it does not involve any Carbon Dioxide (CO_2) emissions. The Hybrid Vehicle Industry have taken an interest in this design to charge the battery.

To give an example of how this machine may work: the Generator would be a round tubular shape with the Magnet fixed on the shaft which moves backwards and forwards with the Piston within a set of Field Coils as in Fig. 43. A Spark Plug forces an explosion with the fuel in the Combustion Chamber, this forces the piston down moving the magnet one way along the Field Coils producing an electrical output.

The Piston forces the shaft with the Magnet attached down and there is a spring chamber that is compressed at the end of the stroke. This is shown in Fig. 43. Although in the diagram the spring has an actual metal spring, in some cases, this may be a Compressed Gas Spring.

There is a system called Homogeneous Charge Compression Ignition where the fuel mix is compressed to the point of auto-ignition and a spark plug is not required like Fuel Injection.

There are three types of Free Piston Linear Generator:

1) The type discussed above in Fig. 43 and Fig, 44, which is simply the Single Piston Type.

2) Next the Opposed Piston type where the Combustion chamber is in the middle and forces a piston either way so it would have two Pistons, two Magnets and two sets of Field Coils. This type is very commonly used in the automotive industry.

3) Finally, the Dual Piston type has a Combustion Chamber at each end, two Pistons, two Magnets etc. and the Field Coils are in the middle.

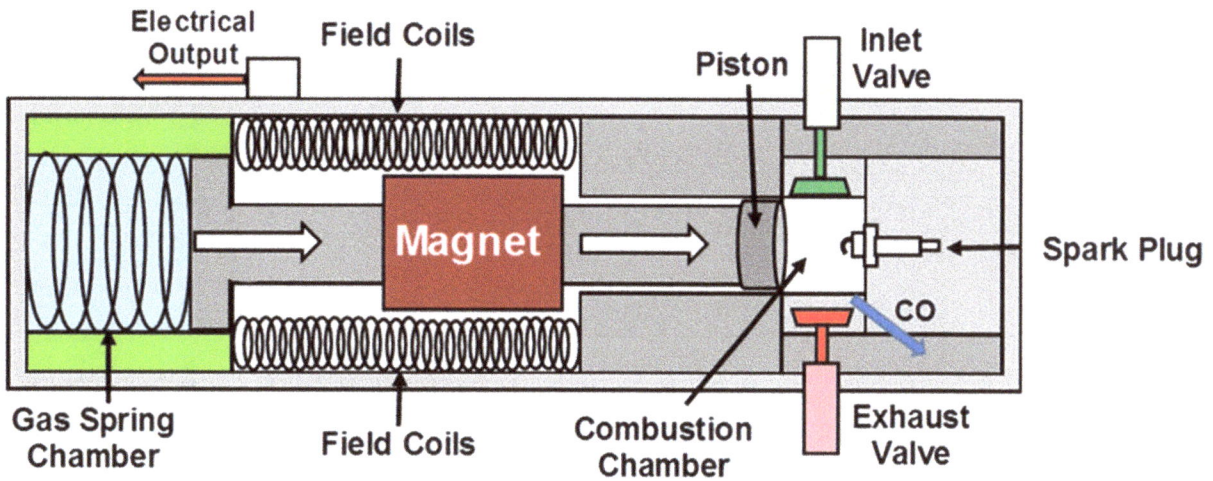

Fig. 44

The Spring Return in Fig. 44 above, shows that after the Combustion Explosion, the Spring System forces the Piston back up the Cylinder and forces Carbon Monoxide (CO) Exhaust Gas out of the Exhaust Valve and the chamber is then ready to receive new fuel and ignition.

At the same time, the Magnet moves in the opposite direction along the Field Coils producing an electrical output. Fig. 44 is drawn as a 4 Stroke Combustion Engine for ease of explanation, whereas the actual Generator Combustion Engine may be 2 Stroke. Having no Crank Shaft involved is a huge advantage as there are no bearings immersed in an Oil Sump to consider.

A conventional spring consists of a coil of wire, usually made of stainless steel or alloy, and because of its makeup it can be compressed and when the compression is taken off, it returns to its original shape. This sort of spring is very common in a whole range of instruments.

A Gas Spring is usually a damper type of unit that uses compressed gas as a sort of gas cushion in a cylinder for equipment that it is used on. This is ideal for the Free Piston Linear Generators as there is no actual physical spring to malfunction or break and the piston is free to have frictionless movement.

What sort of output Voltage can be expected from the FPLG? Around 250-300Volts normally. This type of Generator is very useful for other areas such as the Automobile and the Aircraft Industry.

Can this Generator be installed in a Hazardous Zoned Area? Unfortunately, with its Combustion Chamber it is very doubtful, even with Hydrogen fuel. However, the Hydrogen power would be a Renewable energy.

Brushless Alternator:

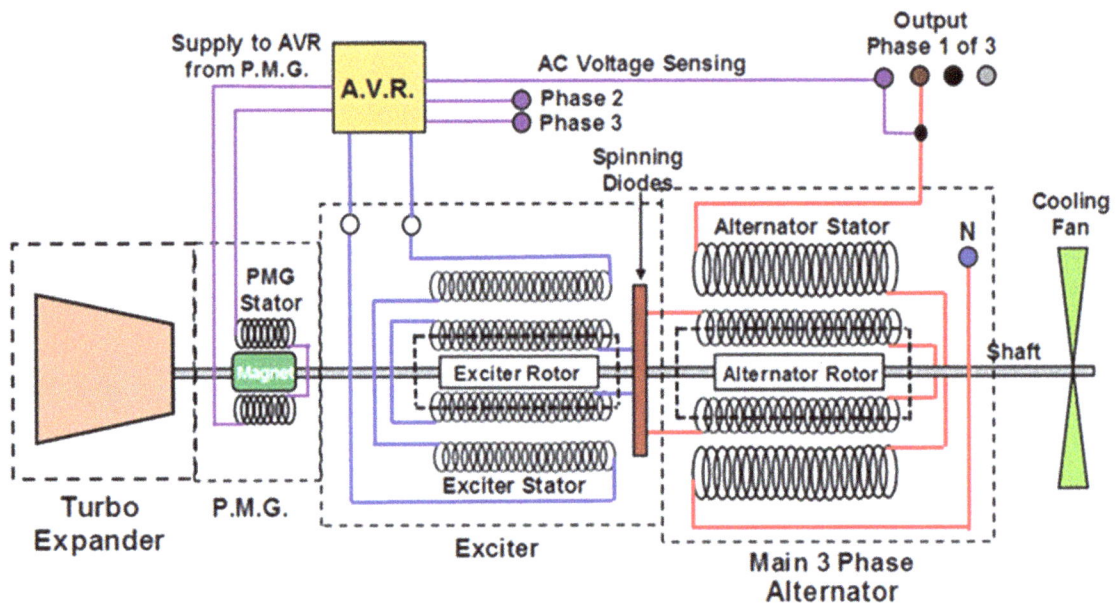

Fig. 45

The Brushless Alternator is ideal for large machines in Hazardous Areas and is Atex approved. This is due to there being no sparking Brushes. The Field Coils in this case are, as in Standard Alternators, a Stationary Armature. So, this is a Magnet (Rotor) spinning in a Coil. An alternative name for these machines without Brushes is a Synchronous Generator and the Alternators with Brushes are called 'Brushed' Alternators.

The Exciter is another DC Generator spinning at the same speed, on the same shaft as the main Alternator so, in this case, with the Exciter, there is a coil spinning in a magnetic field. This DC Generator supplies DC to the main Alternator Rotor to make it into an Electromagnet.

A Standard Generator generates AC, but through a device called a Commutator it converts that AC into DC to make the Main Alternator Rotor into an Electromagnet and that would require sparking Brushes, both on the Exciter Commutator and the Alternator Slip Rings.

The Commutator and Slip Rings are replaced here by Spinning Diodes that convert the Exciter AC to DC to feed the Main Alternator Rotor and they are all spinning on the same shaft. A higher or lower voltage output depends upon winding turns inside of the Alternator and the Density of the Magnetic Flux, which is where the Exciter comes in.

The Main Alternator is generating Three Phase, but in Fig.44 I have shown only one phase as an example. The Field Coils in this case, as in a Standard Alternators, is the Stationary Armature. The output from the Field Windings is Three Phase with the other end of the windings starred into a neutral. The Three Phase Output is taken from the Alternator Stator.

Can the Brushless Alternator be run as a Motor? The answer, from my experience, is **DEFINITELY NOT**. The Circuit Breaker accepting the output will have safeguards fitted to ensure that it cannot be closed on a stationary machine. If the Circuit Breaker did close accidentally, the electrical stresses and strain on the Stator Field Coils would be enormous and the electrical and mechanical forces could cause it to move round in the Yoke, which would be catastrophic.

The Output of a Brushless Alternator is the same as the output of a Standard Alternator. Having worked with the machines, the Brushless Alternator output tends to be smoother, possibly with no friction of a Commutator, Slip Rings or Brushes to contend with.

The Automatic Voltage Regulator (AVR) is still present, as with a Standard Alternator and as mentioned earlier, but just to recap: the simple answer, without going into too much depth, is that the AVR is a piece of equipment in a Alternator Control Circuit that keeps the Output Voltage at a particular pre-set value even if the load was not stable, and prevents surge.

The AVR monitors the Three Phase output and is linked into the excitation, and it will adjust the Field Flux accordingly. It will also monitor Over/Under Voltage & Field Flux Density.

The Permanent Magnet Generator (PMG) is another Generator that comes into the equation here. This generator type runs from a small Gearbox on the same spinning shaft as the Exciter and supplies the AVR itself, with electrical power. This Generator is a Permanent Magnet Rotor spinning in a Stator Coil.

Does the Automatic Voltage Regulator (AVR) need to have a PMG? No, this is done for convenience more than anything. The AVR can get its own supply externally.

What would be the result of the AVR going Faulty or losing the signal from the PMG? This could result in high, low or no output voltage from the Alternator. If the Alternator did not have an AVR at all, there could be problems if the load on the Alternator was to fluctuate or, for some reason, the output required more voltage.

Increasing Rotor Speed can also lead to higher voltage, although this might be the more difficult to achieve and will have an effect on the power frequency (Hz). The Driver in Fig 45 is a Turbo Expander, but this can also be a Gas turbine, Engine, Wind Turbine etc.

A High Voltage Output is an option with Brushless Alternators and can be HV 6.6KV or 3.3KV. A Low Voltage Brushless Alternator is also an option, and this type are fitted in, for example, cars.

Does an Alternator such as this require an Exciter? Technically yes, because the machine still requires the DC for the Alternator Rotor to make it into a magnet. Although this machine can be separately Excited, would it still be Brushless?

The DC must come from somewhere using separate excitation. It would have to have Slip Rings and Brushes in which case it would be called a 'Brushed' Alternator and here we would be back to sparking Brushes on Slip Rings.

It is not the Generator that is termed Renewable or Non-Renewable, more the fuel type driving the machine and as this is usually a Turbo Expander which is part of the process, there would be no problem. If the driver was, say, a Gas Turbine then being natural gas it would be Non-Renewable fuel. If Hydrogen was used, then it would be Renewable fuel.

The Brushless Alternator with a fixed Exciter and Spinning Diodes will be available with Atex approval, to go in a Hazardous Zoned Area. The factory that I worked in had four Brushless Alternators, two of them originating from the 1960s. They were positioned in Zoned Areas, well before Atex came out.

Bottle Bicycle Dynamo:

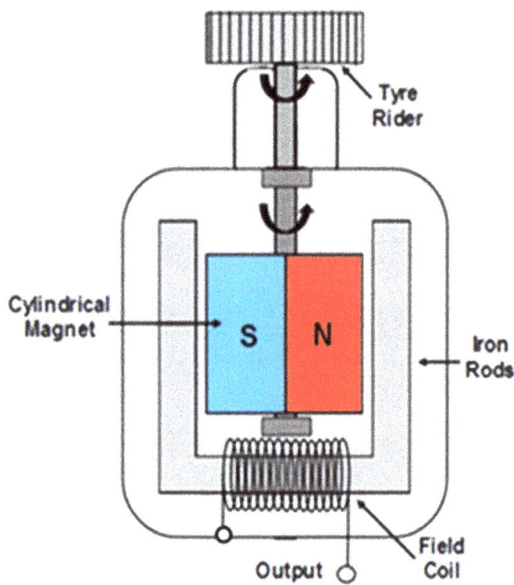

Fig. 46

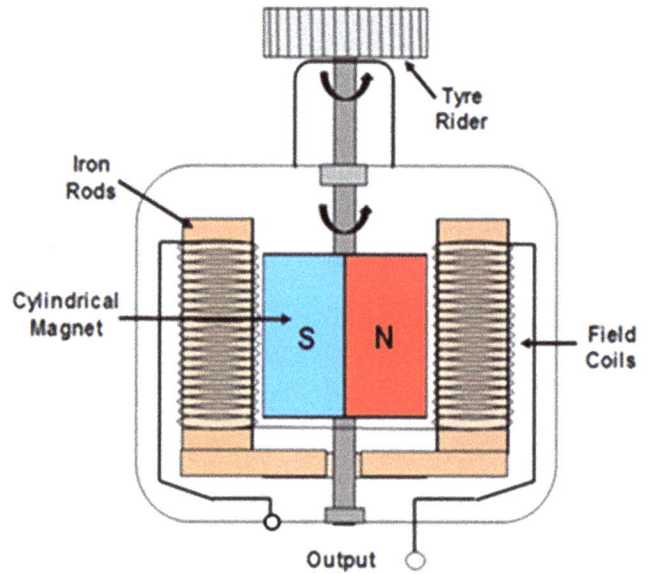

Fig. 47

The **'Bottle'** Dynamo as in Fig. 46 and Fig. 47, is a very small machine compared to the types that we have been talking about up to now. This is one of the most popular Dynamos found in an older Bicycle Lighting System. Another name for these dynamos is a **'Sidewall'** Dynamo and are so-called because they ride on the sidewall of the Bicycle Tyre and are shaped like a Bottle.

There are two slightly different designs. The Field Coil in Fig. 46 is at the bottom of the Iron frame and in Fig. 47, there are two Field Coils around the iron frame at either side of the Permanent Magnet. Hub Dynamos are also a very common Bicycle Dynamo and usually mounted right in the centre of the front wheel, and they work on the same principle, but are much quieter.

Whichever design is chosen, the Rotor is usually a Permanent Magnet spinning in Field Coils, so they are also sometimes called Permanent Magnet Dynamos (PMDs) and are specially for bicycles. This is where Dynamos and Generators become very similar in the way they work.

Bicycle Dynamos such as these are DC, because the Coil Current does not change direction as it spins in the Magnetic Field. A Battery can be included so the Dynamo charges the Battery, however the more that is added will contribute to the weight of the Bicycle.

A Rim Dynamo is the modern-day version of a Sidewall Dynamo and instead of riding on the wall of the tyre, it is made of alloy and has a rubber 'O' ring inserted into a spinning wheel that rides on the bicycle wheel rim just below the tyre. Rim Dynamos are much smaller and do not seem to have the drag of the Sidewall Dynamo.

Will the Bottle Dynamo still work if the tyre is wet? The answer is yes, but not as efficiently. When a Dynamo or Generator is under load, the rotor becomes much harder to turn and the more lights it is feeding, the worse this is and when coupled with the drag on the tyre, it may slip. This could also be the case with the Rim Dynamo.

The only design that would be the best option here is the Hub Dynamo as it is built into the centre of the bicycle wheel (the hub).

Anemometer:

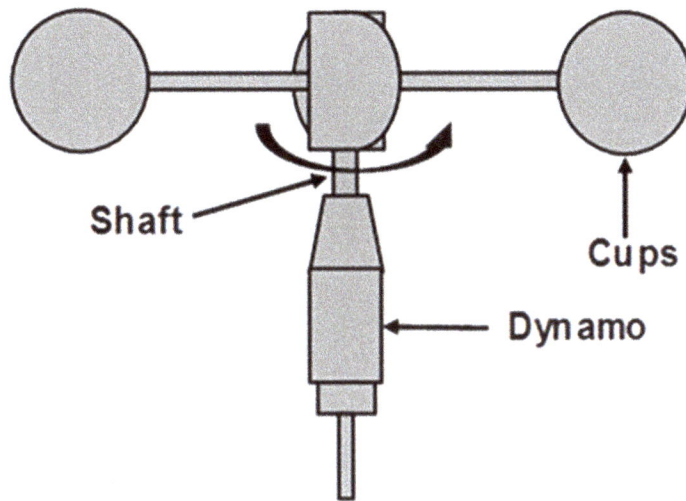

Fig. 48

The Anemometer is the instrument shown in Fig. 48 above. It has cups that are mounted on a shaft and connected to a dynamo, very similar to the bicycle one, and are used to measure wind speed. The Anemometer, with no generator of course, has been around since the 15th century.

The Dynamo type works by using the wind speed to spin the dynamo to create voltage. The higher the wind speed, the faster the cups spin and in turn this spins the Dynamo, which is below the cups, and so produces more volts. The power being produced is fed into a printed circuit which in turn operates an instrument similar to an Electronic Speedometer.

Various Uses:

a) One important use of these machines would be, for example, on the roof of the Control Room on a Chemical Factory or Platform, in case of a gas release and together with the weathervane, they can tell wind direction and speed of the release.

b) Another important use of an Anemometer is on the weathervane on the back of the Nacelle of a Wind Turbine, to send a signal to the Wind Turbine Controller or Processor indicating wind speed, which in the case of a Wind Turbine is extremely important in case the wind is too strong.

c) A Scientific Weather Station would have an Anemometer mainly to just measure the Wind Speed.

d) Certain Cranes. If the wind is too strong, they will not use a High Jibbed Crane for fear of being blown over.

There are different types of Anemometers. Some are like the one in Fig. 48 above, and others which are Fans turning a Dynamo, there are types with Counters and Paddles, which count the revolutions, and some more modern Anemometers operate by Beams of Light. These are called an Optoelectronic/photoelectric system. Other types of modern instruments are even Ultrasonic or Laser.

A Handheld Anemometer can be obtained, but this is a fan turned by the flow of wind rather than using cups. The data goes to separate unit to produce is a digital readout. Although it is possible, it is unlikely for the wind to be too strong for the Handheld Unit.

What is a Magneto?

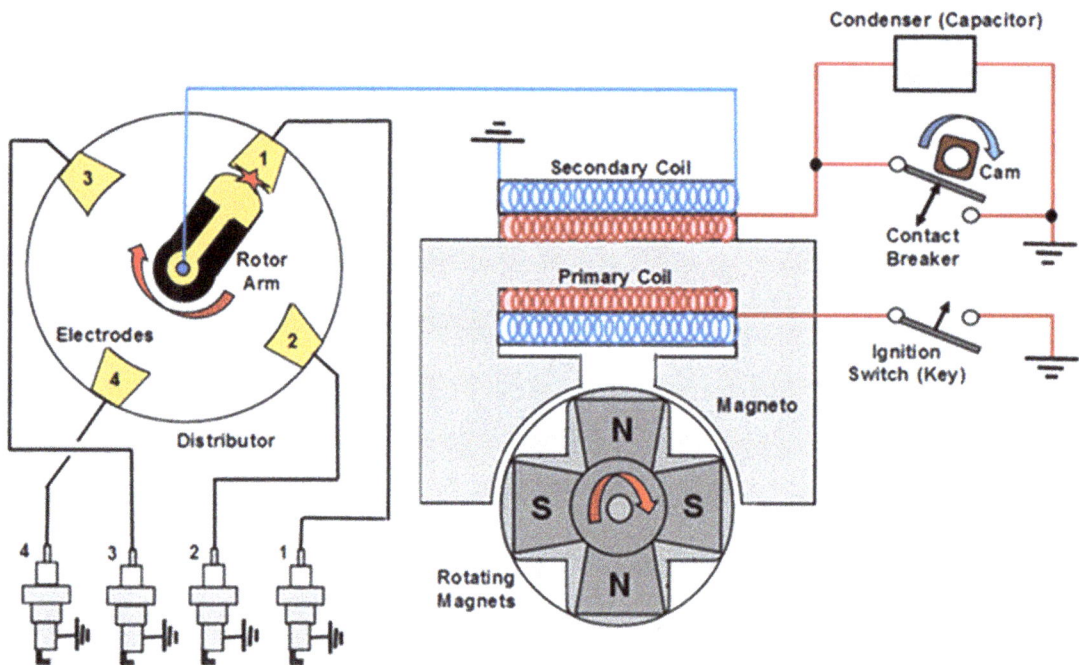

Fig. 49

A Magneto is a type of Generator used as part of some small Ignition Systems in the past. The beauty of this system is that it relies purely on the engine and does not involve a battery in any way, so there is no charging system.

There were three main types:

a) Rotating Magnet as per Fig. 49 above.

b) Rotating Armature: A Coil revolving around stationary Permanent Magnetic Poles.

c) Polar Induction: Both Permanent Magnets & Coils are stationary, and inductors keep oscillating the Magnetic Flux.

The Ignition Switch is turned on and the engine turned over, which in turn causes the cam to turn and open and close the contact breaker as in Fig. 49. The Rotating Magnets on a 'Flywheel' spin with the engine causing the magnetic flux of the Permanent Magnet to cut the coils.

The Contact Breaker is in circuit with the 200 Turns Primary Coil (Red) and the opening and closing action caused by the cam induces a high voltage onto the 20,000 Turns Secondary Coil (Blue).

This High Voltage is transmitted to the rotor arm inside of the distributor, where a Carbon Brush rides on the top centre, and hence HV to the spark plugs is produced by the collapsing field cutting the coils.

System 'Timing' is critical to the firing order to each of the spark plugs, hence the engine runs, and the higher the engine revolutions, the higher the voltage. Remember, it is the points opening and collapsing of the Coil Field that causes the high voltage, which will be several thousand volts.

The Condenser in Fig. 49, in this case, is another name for a Capacitor and is connected in parallel to the Contact Breaker. When the Spinning Cam allows the contacts to open, the Condenser charges up and with the current now flowing into the Condenser instead of the circuit, we eliminate the arc and the current in the primary coil does not drop to complete zero. When the points close, the current flows back into the Coil from the Condenser.

The above action by the Condenser greatly speeds up the collapse of the field and in turn will lead to a higher voltage and a much more efficient system. The voltage is not constant, just a pulse tuned to the engine.

There are different Distributers available: there are the types in the diagram above where there is a rotor arm and a gap. There are also types with a Carbon Brush.

Faults in the past were sometimes nothing to do with the generation. The main problems on these systems were:

a) Faulty Condensers.

b) 'Crusting' caused by arcing on the Distributer Rotor Arm.

c) 'Crusting' caused by arcing on the Electrodes.

d) Electrodes stuck together.

e) Incorrect Spark Plug Gaps. (Gap too wide)

f) Incorrect Spark Plug Gap. (Not wide enough or no Gap)

g) Break in Primary Coil.

h) Break in Secondary Coil.

i) Carbon Brush on Rotor Arm not making contact.

j) Oil on the inside of the Chamber. (Seal gone)

k) Loose Screw connections.

l) Engine Timing out.

Uses for this type of system are:

a) Mopeds.

b) Lawnmowers.

c) Chainsaws.

d) Model Planes.

e) Small Wind Turbines.

The Magneto is not used on cars because the engine must turn over before the Magneto will work, and a Battery must come into the equation somewhere! A very highly sophisticated Magneto is used on light Airplanes, but this is a very advanced machine compared with the ones used on mopeds.

Magnetos produce AC current as there are no Commutators or Diodes in the circuit to change the current to DC. A Battery is not included in the circuitry because the standard Magneto is purely an ignition system. When the engine turns it generates, so no generation or power with the engine stopped.

Frederick Simms (1863-1944) British Mechanical Engineer and Robert Bosch (1861-1942) German Engineer and Inventor together developed the first Magneto around the 1890s.

Types of Generator Drive:

In this section, we will discuss which drive unit turns the Generator; is it Steam, Water, Wind, Combustion or Geothermal? In the 21st Century we are looking at processes that:

a) Are Renewable energy.

b) Do not release CO_2 (Carbon Dioxide.) 1 Atom of Carbon, 2 Atoms of Oxygen.

c) Do not release CO (Carbon Monoxide.) 1 Atom of Carbon & 1 Atom of Oxygen.

d) Are not Nuclear Fission, although this may be staying with us for some time.

Renewable Energy means that it will not run out in the future. So we will look at the later examples of Generator driving processes that are Renewable and Non-Renewable: There are pro's and con's for many of the process fuels even if they are Renewables.

The future, as far as fuels are concerned, is looking good. Below are a few examples that although are experimental now, the way science is going, it will not be long before more advanced systems are developed. Examples of these are:

1) Rainwater Generation.

2) Biomass which uses 'Refuse' instead of Wood Chips.

3) Wind Power without Bladed Turbines.

4) Wind Turbine Power with higher 20MW+ output.

5) Electric Vehicle Batteries that charge in minutes.

6) Tidal Generation in every Major River.

7) Wave Generation just offshore.

8) Floating Photocell Farms.

9) Hydrogen as the Major Fuel for vehicles.

10) Hydrogen powered domestic generators.

11) Nuclear Fusion instead of Fission.

Combustion Engine Driven Generators are with us at present in the form of Vehicle Alternators, Industrial Generators and Domestic Portable Generators. For instance, at the moment, a Domestic Portable Generator may run on Petrol, but Petrol is a 'Non-Renewable' fuel meaning that sometime in the future it will run out. So, what fuel can be used to take its place?

Oil Based Fuels are made from Oil, but Oil itself cannot be artificially made. If companies keep on extracting it out of the ground at the rate that they are doing, then there will come a time when there is just no more to extract so Oil is a 'Non-Renewable' Fuel.

Fossil Fuels produce a great deal of Carbon Monoxide (CO) and Carbon Dioxide (CO_2), the word that is common here is **'Carbon'** all of which is adding to the Global Warming. As with Oil Based fuels, Fossil Fuels will eventually run out, so again making them 'Non-Renewable' Fuels.

Power Stations that use Fossil Fuels are being phased out because of the Carbon Dioxide (CO_2) that they introduce into the atmosphere. The last British Coal Power Station closed in September 2024. Coal of course is **Non-Renewable.**

Why not just use Wind Turbine Generators instead of Power Stations? Well, to replace **ONE** of the largest Power Stations (3GW) would take around **375** Wind Turbines at **8MW** each. It is obvious that if they can substantially improve the Turbine MW to, say, **20MW** Capacity, we would not require as many. Also, these Sea Wind Farms, like the Dogger Bank Wind Farm, can be as many as 80 miles out in the sea, away from our shores.

Gas Power Stations, apart from Nuclear Fission Power Stations, are our main Fuel now, but Gas of course is **Non-Renewable** fuel. To replace all of our Gas Power Stations **(28.68GW or 28,680MW)** would take around **3,585** Wind Turbines at **8MW** each. So, if you think that there are a lot now…

Questions are asked, "Why don't we build all Wind Turbines out in the sea?" A Fixed Sea Wind Turbine costs around **£4m/MW** of electricity producing capacity, against **£1m/MW** of electricity producing capacity building them on land. Sea Wind Farms can only be built in shallow sea water, up to, say, **60 Metres.** There is the option of Floating Wind Farms which are discussed in more detail later in the book, but these have to cope with rough seas and storms etc.

Whichever option is chosen, Sea Wind Farms are very expensive and the one problem that is always overlooked by the public, is the fact that from each Wind Turbine there is a **HV Cable** to cater for, sometimes going to 'Hubs' (Substations) out in the sea. There would usually be one/two 'Hubs' per Sea Wind Farm depending upon the number of Wind Turbines in the Farm, otherwise all the HV cables from all of the Turbines would have to come ashore to a substation on land.

So, there are many arguments put forward by the Government about why we build so many Wind Farms on land, such as cost and targets promised by governments which must be met.

These Non-Renewable Power Station Fuels below, as mentioned previously, will run out in the future:

1) Coal –   Fossil Fuel – CO_2

2) Gas –   Fossil Fuel – CO_2

3) Oil –   Fossil Fuel – CO_2

4) Diesel –   Made from Oil.

5) Emulsified Diesel – Made from Oil

6) Petrol –   Made from Oil.

7) Methane –   Natural Gas.

8) L.P.G. –   **L**iquefied **P**etroleum **G**as: Propane.

9) L.P.G. –   **L**iquefied **P**etroleum **G**as: Butane.

10) L.P.G. –   **L**iquefied **P**etroleum **G**as: Isobutane.

11) L.P.G. –   **L**iquefied **P**etroleum **G**as: Contains Propylene.

12) L.P.G. –   **L**iquefied **P**etroleum **G**as: Contains Butylene.

13) L.P.G. –   **L**iquefied **P**etroleum **G**as: Contains Isobutene.

Unfortunately, our Wind Farms, Solar Cell Powered Farms, Wave Power and Tidal Power produce nowhere near as much energy that we require, so currently these are just used to top up the existing Power Stations at the. It might be a different story by 2030.

Renewable future Power Fuels are constantly being invented, some examples of electricity generation of the future are as follows. I am sure that more can be found:

1) Nuclear Fusion –   Experimental at the moment.

2) Nuclear Fission –   What we have at present and well into the future.

3) Wind Turbines –   Horizontal that we have at present.

4) Wind Turbines –   Vertical Turbines.

5) Wind Turbines –   Bladeless.

6) Wind Turbines –   Fixed to seabed – Jacket.

7) Wind Turbines –   Fixed to seabed – Mono-pile.

8) Wind turbines –   Fixed to seabed – Gravity.

9) Wind Turbines –   Floating – Spar Buoy.

10) Wind Turbines –   Semi-Submersible.

11) Wave Power –   Buoys.

12) Wave Power –   Rods.

13) Wave Power –   Attenuator.

14) Wave Power –   Compression.

15) Wave Power –   Overtopping.

16) Tidal Power –   Fixed.

17) Tidal Power –   Barrage.

18) Solar Power –   Rooftop.

19) Solar Power –   Portable Generator Land Based.

20) Solar Power –   Floating.

21) Hydrogen Fuel Cell –   For Industrial/Domestic use.

22) Bio Diesel –   Produced from Vegetable Oil and Fats.

23) Bio Fuel –   Wood Chip.

24) Bio Fuel –   Produced from Corn & Sugarcane.

25) Hydrogen –   Most common Element ever.

26) Nitrogen –   Used for its pressure not combustion (Rare fuel)

27) Ethanol –   Produced from Alcohol.

28) Geothermal –   UK Underground Volcanic Action (quite rare).

Many of the above systems will not be in our Hazardous Areas, but they are familiar to us, so I have included them for interest.

As already mentioned, Renewable Fuelled Generators will have to increase their outputs from 7 – 10 Mega-Watts (MW) to around 20 Mega-Watts (MW) to be serious contenders to replace Power Stations of the future!

Atmospheric or Natural Draft Cooling Towers:

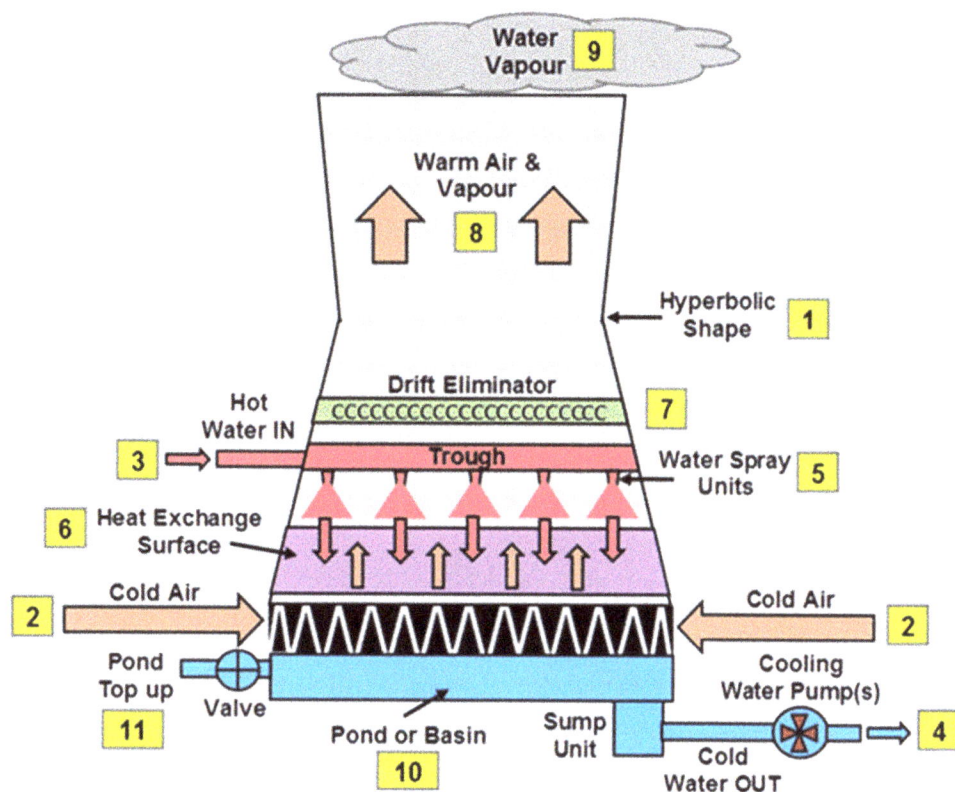

Fig. 50

Atmospheric or Natural Draft Cooling Towers are the first noticeable structures when we see Power Stations. This is not electrical, but an explanation of how the Atmospheric/Natural Draft Cooling Towers work may be advantageous. Basically, they cool the water from the plant, as the radiator in your car cools the water heated by the engine. See Fig.50.

For anyone (authorised) wanting to enter the Cooling Tower Internal, access to the inside of the tower would be by means of a set of concrete steps leading up to a door which would be on the level of the Hot Water in (3. For instance, where I used to work, Technicians used to enter the Cooling Tower whilst the tower was still in operation and change the spray units.

The Shape of the tower is very special. Firstly, the tower is tall, they can be around 250-300 foot tall, so that Water Vapour does not come down to plant level and cause disruption. Large volumes of air and water vapour rises inside of the tower (8) and is compacted by the 'Hyperbolic' shape (1) and then released out of the top with huge acceleration, causing a tremendous updraft, which sucks cold air (2) into the tower at the bottom.

Water Vapour is seen coming out of the top (9) which is carried up by the large volume of air and this can add up to hundreds of gallons of water, as vapour, exiting the tower each week. There has to be a top up system (11) to top up the basin or pond (10), as it can be called upon to cater for the amount of water vapour coming out of the top (9).

*Note: This is 'Water Vapour' which is **NOT** the same as **Steam.***

The Top-up Water which tops up the basin, can come from two sources. It can come from the towns water supply, or if the Cooling Tower is located next to a fairly large river, it can come from there, although there will have to be a filtering system, and it will have to be treated.

The Drift Eliminator (7) prevents quite an amount of Water Vapour from going up the Cooling Tower with the air draft and reflecting it back down through the heat exchanger. Between the 1960s – 1990s there was no such thing as a Drift Eliminator, so there was more Water Vapour coming out of the top.

The Hot Water (3) enters the tower around a third of the way up and runs along troughs inside of the tower and out to the spray units (5), to be sprayed through the heat exchanging material (6). These days the heat exchanging material is plastic sheets of 'Hexagonal Shapes' to give the maximum surface area. The falling hot water is met by the cold air draft being sucked in at the base of the tower, through the heat exchanger and up towards the top.

On older cooling towers this heat exchanging material would be 'Wood Slats', some of which had to be replaced yearly as it broke, especially if they iced up due to the weight of ice.

The amount of water in the System at any one time, including the Basin or Pond (10), could be up to 1,000,000 gallons. This makes this water ideal as a standby in case there is a fire. A Fire Pump could be connected to the basin for emergency water. This was an invaluable source of many thousands of gallons water for the Fire Pump on the factory where I worked.

In the Basin there is a sump where water is extracted by Cooling Water Pumps (4) and pumped back to the Condenser base (Fig. 50). It would not be unusual on a large Power Station or Chemical Plant to have several High Voltage (HV) Cooling Water Pumps.

Freezing Up in winter can cause the Cooling Tower real problems. Again, back in the past, the Cooling Tower could freeze up in winter causing large icicles to form on the Wooden Heat Exchanger slats and the weight did cause some of them to break. A Defrosting Ring using diverted warm water ran in a circular pipe around the base of the tower to try and prevent this. I can assure you that this never really worked very efficiently!

Tower Inspections must be carried out every few years. The concrete of the tower has to undergo this inspection, and steeplejacks will bolt a ladder to the concrete from the base to the top and climb up to inspect the concrete at the top. Just pause for a moment and think about this ladder taking the same hyperbolic shape as the Tower and someone climbing up it and walking round the rim. Obviously, the Tower will be shut down whist this Inspection is going on.

Water Treatment of the Cooling Water will also be carried out by a Water Treatment Plant and treated for things like:

a) Alkalinity: Can cause corrosion.

b) Acidity: Can cause corrosion.

c) Chlorides: Can be corrosive.

d) Hardness: Contributes to scale.

e) Iron: With Phosphate can lead to blocking.

f) Organic Matter: Can lead to blocking

g) Silica: Can cause scale.

h) Sulphates: Corrosive to metal.

One inspection that is very important to be carried out on Cooling Towers, or wherever there may be Evaporative Cooling Systems concerned with water, is a **Legionella Bacteria Test** to keep employees safe.

Wet Surface Cooling Towers

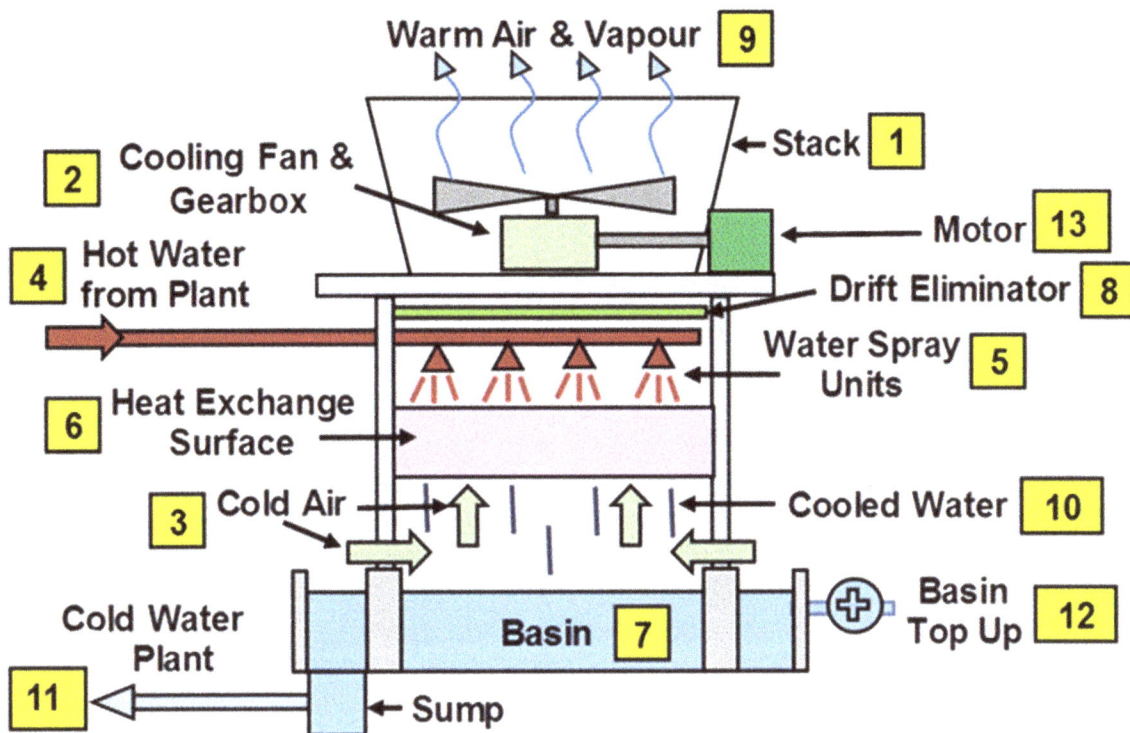

Fig. 51

Wet Fan Cooling Towers are slowly replacing the Natural Draft ones. Again, this is not exactly electrical in the sense of what we have been discussing, but being so common it will be advantageous to know how these Cooling Towers operate.

Cooling Towers of today are probably of the kind shown in Fig. 51 above. The Concrete Cooling Towers of the past are very permanent and costly to demolish should the plant become redundant. Maintenance such as changing the spray units when required must be done and then there is of course an electric motor, gearbox and fan unit.

Open Loop Cooling Towers are common and are the same as in Fig. 51 where the water enters the tower near to the top and trickles down through the heat exchanger into the Basin.

Closed Loop Cooling Towers look similar to Open Loop Cooling Towers, except that the plant cooling water never comes into contact with the Tower Basin. The water remains in pipes in a closed loop, which pass through the tower and cold water drops onto the pipes cooling the water inside, which is constantly being pumped round the loop.

Electricity is one thing this type of Cooling Tower will use a great deal of, which the Natural Draft ones of course do not. The cooling fan electric motor can be running 24/7 and will use a vast amount of electrical energy.

The Cold Air System works as follows: Firstly, there is the stack (1) which houses the gearbox and huge fan (2) fed by a Motor (13) which is outside of the Stack. As the fan spins, it drags a huge amount of cold air in at the bottom (3).

The Hot Water from the Plant (4) enters the Cooling Tower, at around 80ºC+, and proceeds through the Water Spray Units (5). These units spray the hot water onto the Heat Exchange surface (6) and down into the Basin (7).

The Cold Air Intake (3) blowing through the water droplets, below and up into the Heat Exchanger (6), cools the water (10) on its way up towards the Stack (1), with the air being dragged upwards by the huge Cooling Fan (2) and out of the top of the Cooling Tower as Warm Air and Water Vapour (9).

Note: This is 'Water Vapour' which is **NOT** the same as **Steam.**

The Drift Eliminator (8) will return a lot of the Water Vapour back to the basin (7). Wooden Cooling Towers of old did not have a Drift Eliminator, hence more Water Vapour would come out of the top.

The Heat Exchanging Material is Plastic Sheets of Hexagonal Shapes to give the maximum surface area and the falling hot water is met by the cold air draft being sucked in at the base of the tower, through the heat exchanger and up towards the top.

The Cooled Water (11) is returned to plant from the basin (7), via a Sump, by Cooling Water Pumps which will be very powerful, could be, HV or MV Motors.

Basin Top up is done from a water feed float unit controlling, usually what is called Town Water (12), to make up for the vapours coming out the top of the Cooling Tower which add up to hundreds of gallons of water lost. The Cooling Tower Basin can be topped up from a river if one is local, but water must be filtered and treated.

Vapours at Ground Level are not unusual for this type of tower depending upon atmospheric conditions. Remember this is purely Water Vapour and not Steam! I have known it be like a fog with the right atmospheric conditions.

Water Treatment of the Cooling Water will also be carried out by a Water Treatment Plant and treated for things like:

a) Alkalinity: Can cause corrosion.

b) Acidity: Can cause corrosion.

c) Chlorides: Can be corrosive.

d) Hardness: Contributes to scale.

e) Iron: With Phosphate can lead to blocking.

f) Organic Matter: Can lead to blocking.

g) Silica: Can cause scale.

h) Sulphates: Corrosive to metal.

Wet Fan Cooling Tower Construction in the 1960s & 1970s for towers such as these, were made of wood along with the Heat Exchanger section.

Are the Fan Motors High Voltage? They can be HV, but usually they are not, they are 400-415 Volts motors taken through a gearbox. These MV Motors are usually very accessible for easy maintenance. There is a stairway up to the motor floor and it safe to go up with the fan running inside of the stack.

One inspection that is very important to be carried out on Cooling Towers or wherever there may be Evaporative Cooling Systems concerned with water, is a **Legionella Bacteria Test** to keep employees safe.

Typical Power Station:

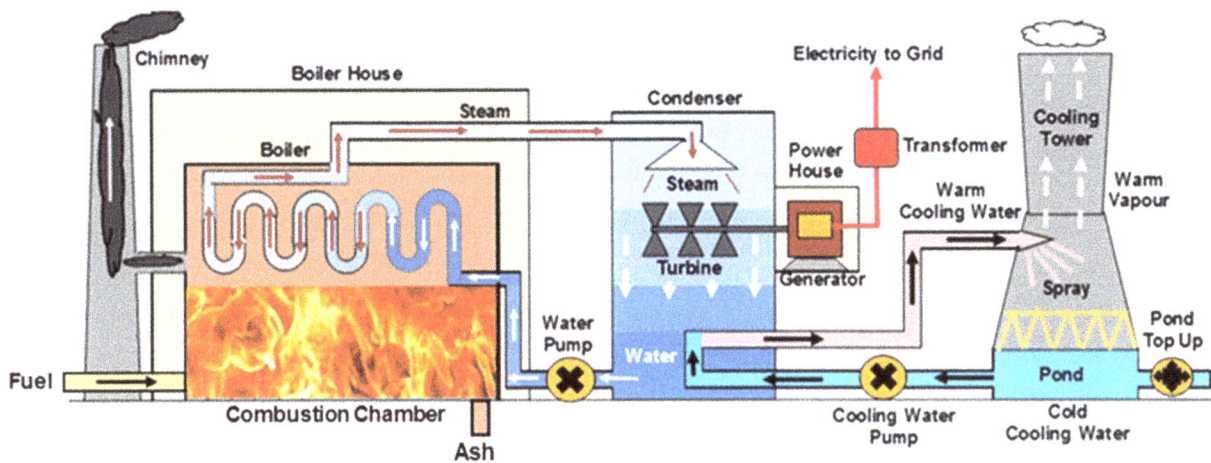

Fig. 52

A typical Power Station process set-up is where a boiler heats water to produce high pressure (HP) steam in the boiler tubes, leading to a condenser to drive a turbine and in turn a Generator as in Fig.52. A bit like your kettle boiling at home on a much larger scale.

The Fuel in this type of process can be Coal, Gas, Oil, Nuclear or Biomass (Wood Chip), although the last Coal Power Station, Ratcliffe-on-Soar, was closed in September 2024. Coal Power Stations were very inefficient, only about a quarter to a half of the energy in coal is converted to electricity, the rest goes on to make unwanted by-products like CO_2 etc.

There are not many Oil Power Stations left, as most stations these days are Gas and some are slowly changing to Biomass. Now, Biomass Wood Chips are not as innocent as they seem as they still give off CO_2, but Wood Chip is not the only Biomass Fuel as you can see below.

The process is that the Chemical Energy stored within the fuel is changed into Thermal, Mechanical and Electrical Energy as an end product. All of the fuels mentioned in this section, except Biomass and Nuclear, are **Non-Renewable** energy as they can run out in the future.

These fuels, except Nuclear, but including Biomass (Wood Chip), all produce vast amounts of CO_2 which of course is one of the greenhouse gases.

Below research has shown the amount of gas fired power stations in the UK:

a) 2 x Gas Fired Power Stations in Wales.   3.58 Gigawatts

b) 1 x Gas Fired Power Station in Scotland.   1.18 Gigawatts

c) 2 x Gas Fired Power Stations in Northern Ireland. 1.03 Gigawatts

d) 24 x Gas Fired Power Stations in England.   22.89 Gigawatts

Twenty-nine UK gas fired power stations generating a total of around: 28.68 GW (Gigawatts)

The largest ever Power Plant in the UK produces around 4 Gigawatts (4000 Megawatts) whilst the largest power station in the world is in China, and is called Datang Tucketuo, and produces around 6+ Gigawatts (6000+ Megawatts).

With Wind Turbines, we currently cannot build enough Wind Turbines to replace all the Power Stations. This is discussed in more detail later in this book.

Biomass Fuels are not only Wood Chip, but they can also be any of the substances below:

a) Small wood pellets imported from Canada & America.

b) Crops such as maize.

c) Sunflower pellets.

d) Olive and peanut shells.

According to research, there are 226 Biomass Power Stations in the UK. If they could get these stations to safely burn a lot of our refuse, what a breakthrough that would be.

Storage of Wood Pellets must be carried out under very strict conditions in open storage units. There are certain storage conditions that must be met. Stored Wood Pellets can give off Carbon Monoxide (CO) and anyone entering a storage facility should wear a CO monitor. Storage giving off CO depends on certain parameters:

a) The Age of the Wood Chip.

b) Storage Temperature.

c) Type of Wood Chip.

d) Dryness of the storage facility.

The Chimney at a Power Station can be 850 feet tall and 85 feet wide. As with all the fuels mentioned, there are waste products, such as smoke, so there must be a Chimney. Power Station Chimneys are made of re-enforced concrete and lined on the inside with Polymer. They say that a high chimney being so high has a better chance of splitting up the outage into a wider area to disperse it more efficiently.

The Boiler is the process of using its Combustion Chamber to heat water inside tubes to make steam on a massive scale and to high boiling points and pressures to drive a turbine as in Fig. 52.

Because the Water Pump is constantly pumping cooled water into the boiler steam tubes and is then being heated by the boiler and turned into HP steam, the HP steam flows out of the top to the condenser and is used to drive the Turbine and hence the Generator.

Condensed Steam after it has been through and turned the Turbine, changes back into water droplets and drops to the bottom of the Condenser and this process starts all over again.

The Water at the base of the Condenser is constantly being cooled by cooled water from the Cooling Tower. For an atmospheric or natural draft and electric fan cooling towers. See Fig. 50, 51 & 52.

Can we eliminate the CO_2 being produced? Well, to a certain extent, they can recover the CO_2 and store it in huge caverns under the ocean, which once held oil, but this must be a finite process and cannot go on for ever. The question now must be, what do we do with it once we have filled the caverns?

Nuclear Fission Power:

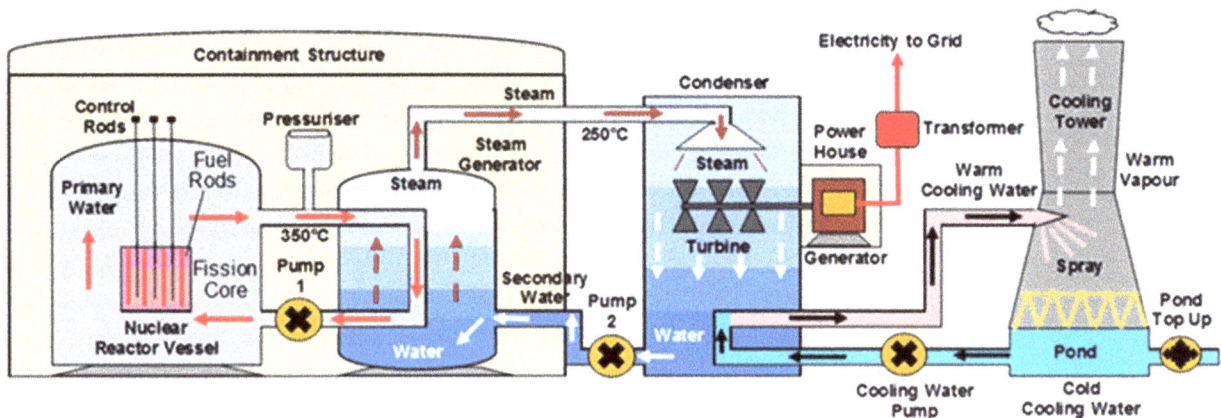

Fig. 53

Nuclear Fission Power Stations are what we have at present, and some people think that atomic energy is fed in at one end and electricity comes out the other. Looking at it from one angle, that is nearly what happens, except that there is a large process between the two. Let us look at each individual stage in Fig. 53 above:

The Nuclear Fission Reactor is housed in a concrete building called a Containment Structure. Inside the Nuclear Reactor, nuclear fission is taking place causing an enormous amount of energy to heat the water in a closed loop, which in turn heats water in the Steam Generator to form steam and in turn spins a turbine. The Nuclear Reactor is carrying out the same job as a Boiler in a Gas Power Station.

The Nuclear Fuel is Uranium or Plutonium. Inside of the reactor, a nuclear reaction is taking place whereby Neutrons are absorbed by the atomic fuel causing atoms to split and release vast amounts of energy i.e. Fission and heat. Looking at how many atoms there are in an element, this amounts to a huge amount of energy and the start of an enormous chain reaction.

Control Rods, containing Neutron Neutralisers, are pushed down into the reactor core and reduce the Neutrons which will slow down the chain reaction, otherwise it could get out of control and possibly cause a meltdown. This is where a reaction getting out of control releases an enormous amount of heat which melts the reactor.

This could also be achieved by losing the coolant, which has two loops in case of failure. Apparently, the Reactor Coolant was lost in the case of Chernobyl. Pulling out all the Control Rods at the same time would have the same effect.

After going through the Turbine, the Water in the base of the Condenser is then cooled by cold water from the Cooling Tower and pumped back by Pump Number 2 to the base of the steam generator, and the process starts all over again. The Cooling Tower can be Natural Draft or Wet Electric Cooling Towers, see Fig. 50 and Fig. 51.

Nuclear Fission Power Stations do not emit CO_2 but there are other bi-products here, such as depleted nuclear fuel which has to be removed and stored.

The question is constantly asked, "can we shut down all Nuclear Fission Power Stations?" The answer is **CERTAINLY NOT!** They produce about 16% - 20% of UK energy, and homes would be in the dark if we shut them down.

Are these Power Stations safe, and have there been any Nuclear Fission Accidents? I believe that they are as safe as they can be, but a few accidents have happened, three examples you might recognise as follows:

a) Sellafield   (UK)   1957

b) Chernobyl   (Ukraine)   1966

c) Fukushima   (Japan)   2011

How many Nuclear Power Plants in there in the UK? According to research, the UK has 11 Nuclear Fission Reactors at various sites. These are advanced Gas Cooled Reactors (AGR):

1) Hartlepool   Durham   England

2) Heysham   Lancashire   England

3) Hinkley Point   Somerset   England

4) Sizewell   Suffolk   England

5) Torness   East Lothian   Scotland

There are more Nuclear Fission Power Plants in the pipeline. They do take a considerable time to build.

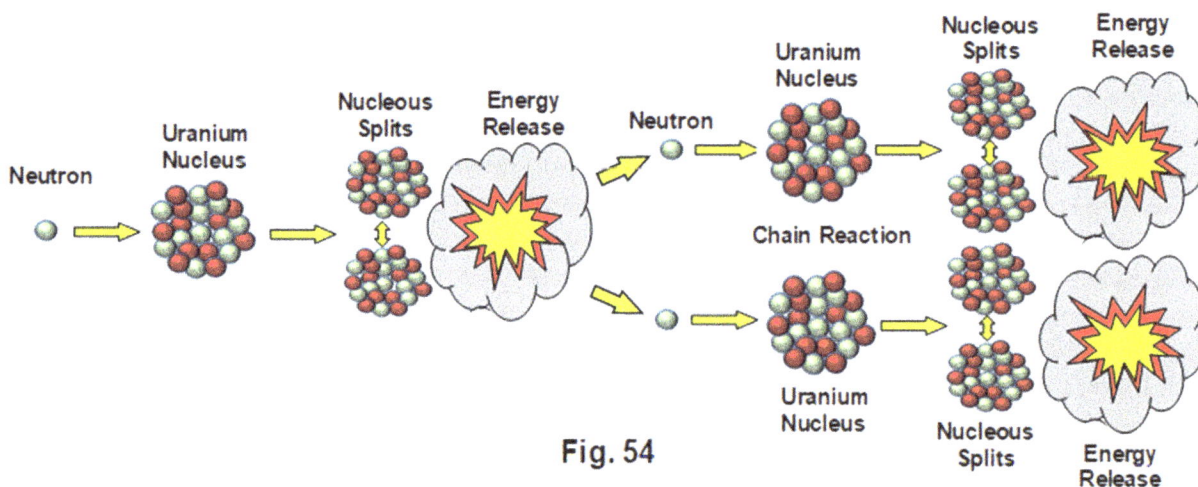

Fig. 54

The Nuclear Fission Reaction is when the nucleus of an atom is split into two equal nuclei by forcing it to absorb Neutrons. So, bombarding the Uranium or Plutonium Element with Neutrons causes the Nucleus to split into two, a very large amount of energy is released at this point causing a Chain Reaction as in Fig. 54. So, as we have mentioned, if we push Control Rods into the Reactor, which are Neutron Neutralisers, we can slow down the nuclear reaction.

If we did not control it, then the result would be an Atomic Bomb. I think you can understand now how to control the Nuclear Reaction. Nuclear Fission was discovered in 1938 by Otto Hann and Fritz Strassmann, who were German Scientists, and is Renewable Energy. For the time being Nuclear Fission is here to stay.

On the news recently, there was a report stating, that in 2-3 years, two of our Nuclear Power Stations will have to shut down, as they have come to the end of their life. This will seriously have an impact on the amount of energy we produce.

Nuclear Fusion Power:

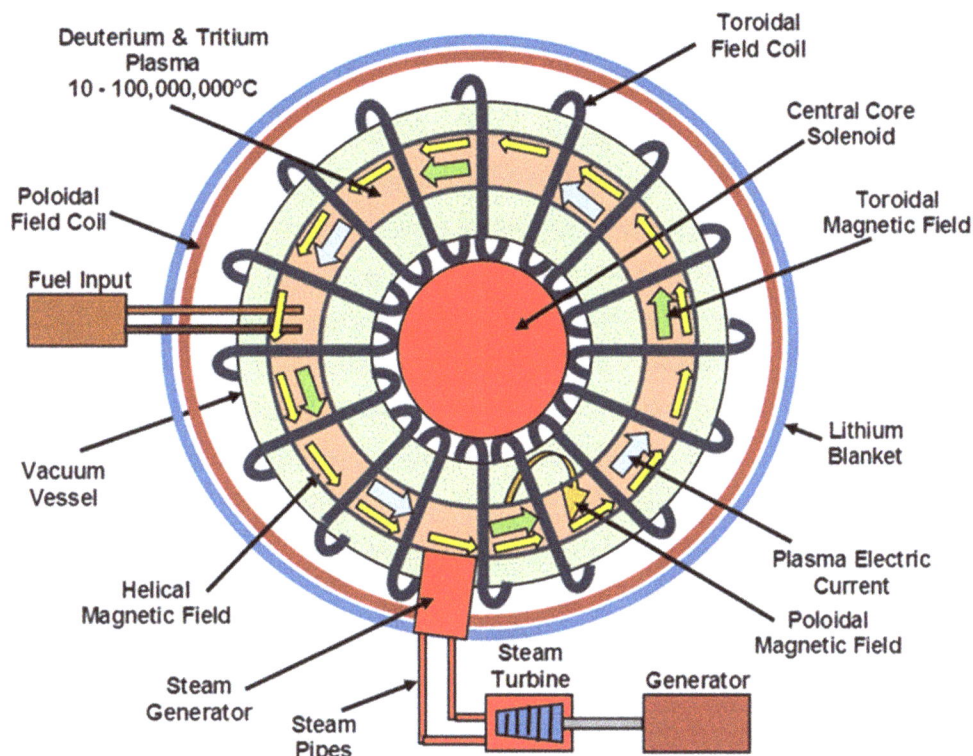

Fig. 55

Nuclear Fusion will certainly not be in any of our hazardous areas at the moment! The actual system of nuclear fusion is very complex. A very basic diagram is in Fig. 55 above showing a Fusion Reactor design.

The word 'Nuclear' usually means that it involves the nucleus of the atom, and 'Fusion' means the joining together. The Fusion Process is an Exothermic Process and is Renewable Energy. The mixed reception by the public to the word 'Nuclear' means that this it will be difficult to sell the idea of this form of energy.

A Nuclear Reaction is where two Hydrogen Nuclei join, instead of 'Splitting' with Fission where a tremendous amount of heat involved. A good example of a huge Fusion Reaction is the Sun! This is a huge Nuclear Fusion Reaction, and the result is Helium and Energy. There are several different designs of Fusion power depending upon the country.

Nuclear Fusion is a safe fuel of the future, not to get mixed up with Nuclear Fission, which is what our nuclear Power Stations are at present. Fusion is achieved when two light atomic nuclei, **Deuterium** and **Tritium,** which are derived from Hydrogen combine to form Helium and Energy.

The enormous heat, which may be in the region of **10,000,000ºC - 100,000,000ºC,** allows the nuclei to collide together and fuse to release huge masses of energy. The reaction causes plasma, which is surrounded by several types of magnetic field.

Waste Products from the fusion reaction are almost Zero and this makes it very attractive. What nuclear waste there is, decays very quickly. There would be no fear of meltdowns and no CO_2. So, as far as that goes, nuclear fusion is very environmentally friendly.

63

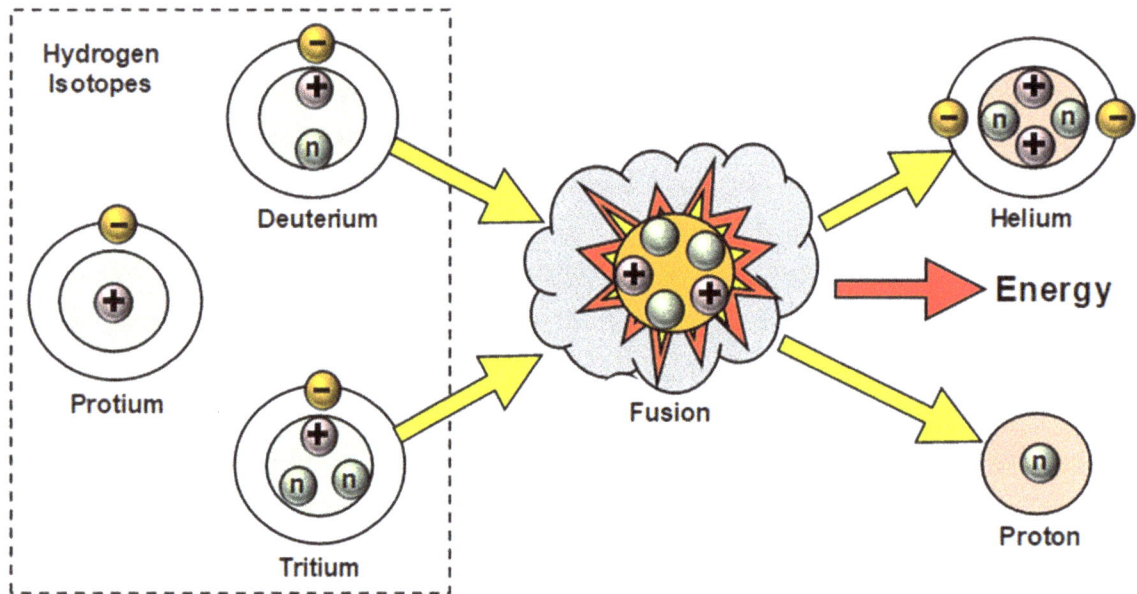

Fig. 56

The Fusion Reaction shown in Fig. 56 above, shows the reaction of two light Hydrogen Isotopes - Deuterium and Tritium, fusing together to make a heavier one, Helium, a spare Proton and a large amount of energy. Protium is also a Hydrogen Isotope but not used in this reaction

Deuterium (D) or (^{2}H) is an isotope of Hydrogen. The largest supply of this isotope is from the ocean. As in the diagram above Deuterium has one Neutron, one Proton and one Electron.

Tritium (T) or (^{3}H) is a bit more complex. Tritium was discovered in the 1930s by Ernest Rutherford and is very rare with only trace quantities found naturally. It is a radioactive isotope of Hydrogen, but it can be made artificially by irradiating Lithium. As in the diagram above Tritium has two Neutrons, one Proton and one Electron.

History is quite recent in relative time! The idea of nuclear fusion is not new, it has been around since the 1930s and discovered by a Chemist called H. Urey.

Currently we are only at the Design Stage of Nuclear Fusion Reactors, but we not far away from a reactor plant, and we could end up with limitless energy, with very little fuel which will last many years using materials that are easily obtainable from Hydrogen.

Can they build a Fusion Power Plant now? No, building a container called a **Tokamac, t**hat can withstand 100,000,000ºC will the biggest challenge.

The Reaction of Nuclear Fusion is hotter that the core of the sun, and science dictated a way to hold the plasma in which was a Magnetic Field, similar to what they have to do with anti-matter, and to this end the Tokamac is lined with very strong magnets.

Future: Even at the stage where science is at this present time, it will take several years to perfect a commercial process that is workable. All I can say is 'watch this space' as this is definitely a fuel of the future.

Note: An Isotope, such as Deuterium and Tritium, are Atoms of the same Element, in this case Hydrogen, and must have the same number if Protons, but may have a different number of Neutrons as can be seen in Fig. 56 above.

Open Cycle Gas Turbine:

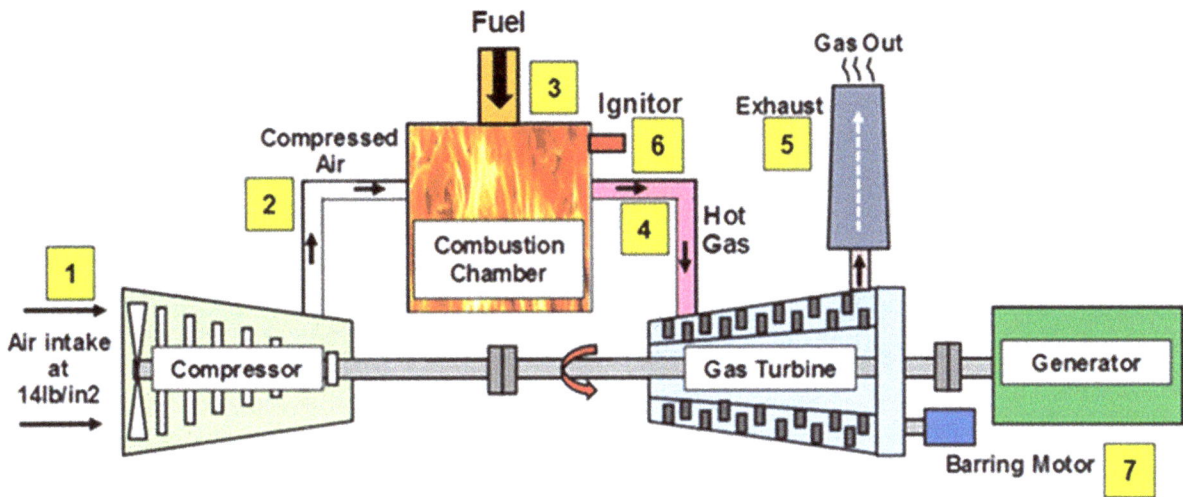

Fig. 57

Open Cycle Gas Turbine as in Fig. 57, makes a Gas Turbine Drive look simple. The Gas Turbines back in the 60s & 70s that I worked on, were so large that they had to have a Resistance Soft Start, HV, 6.6KV Starter Motor to get the Turbine Rotor up to a certain speed before the combustion took over.

These days though, the machines are smaller and much different, but the theory remains the same. These machines have an extremely fast start up time compared with other types of drive.

How do they work? Firstly, there are four main parts:

a) The Compressor.

b) The Combustion Chamber.

c) The Gas Turbine.

d) The Generator.

The Compressor, Gas Turbine and Generator are all on the same shaft. Let us have a look at them one by one. The main objective is to drive a Generator in this case.

The Compressor is required to supply compressed air to the Combustion Chamber at a set, constant pressure. Three other things will be required for combustion:

a) Compressed Air. (1)

b) Fuel which will be Methane (Natural Gas) (3)

c) Ignitor (6)

The Compressor takes a huge intake of air at 14lb/in² (Atmospheric Pressure), which is then filtered, compressed and exits the compressor at very high speed, marked 1 & 2 in Fig. 60 above.

The Combustion Chamber receives compressed filtered air from the compressor, which enters at very high speed and pressure. Methane, which is a Natural Gas, is then fed into the Combustion Chamber, which is marked 3 in Fig. 57, and using a Burner Igniter, marked 6 in Fig. 57, it starts off the combustion process.

After Combustion the hot gas, at over 1000°C, leaves the combustion chamber, marked 4 in Fig. 57, and heads towards the Gas Turbine itself under a very high temperature and pressure. The problem here is that greenhouse gases are released because of the combustion chamber process.

The Gas Turbine has a series of specially shaped blades fitted to turn when the hot gas from the Combustion Chamber is fed onto them and expands. There are **'Fixed'** and **'Moving'** blades inside the gas turbine that miss each other by thousandths of an inch.

The Expanded, Combusted Gas moves through the fixed and moving blades and causes the rotor to turn. When the turbine is at full speed, which could be many thousand RPM, the blades could be at a nearly red heat.

Heat in some very large turbines could cause the shaft to sag if they just stopped the turbine. A Barring Motor, No. 6 in Fig. 57, engages to turn the shaft round slowly until it cools down should it be needed in this design.

The Exhaust Stack is required for the hot gas to exit after it has been through the turbine. It is marked 5 in Fig. 57. The whole system of air intake to exhaust is called a 'Brayton Cycle' after George Brayton, a past developer of the system.

Two different Turbine Systems: (Both Non-Renewable)

a) OCGT – **O**pen **C**ycle **G**as **T**urbines (Shown in Fig. 57 above.)

b) CCTG – **C**ombined **C**ycle **G**as **T**urbines. (Shown in Fig. 58 in next section.)

The Open Cycle Gas Turbine is less efficient than the Combined Cycle Gas Turbine, as no heat is recovered to turn water to steam, which drives a second turbine, but will be much less expensive. This is a Non-Renewable system because of the Combustion Chamber Fuel.

If Hydrogen could be used as a fuel in the Combustion Chamber, then this would become Renewable energy.

The Gas Turbines are used on Aviation (Jet Aircraft), Locomotives (Trains), Marine (Ships), Production of Electricity (Generators/Alternators) and Industrial (Driving Compressors).

The Open Cycle Gas Turbine (OCGT) is one that we are discussing, so is there a Closed Cycle Gas Turbine? In fact, **YES,** there is such a system as a **Closed Cycle Gas Turbine** where the exhaust gas is continually passed back through the system instead of being exhausted to atmosphere.

A Closed Cycle Gas Turbine is **NOT** the same as the Combined Cycle Gas Turbine (CCGT) they are two different systems.

We must ask the question "Are there any working Open Cycle Gas Turbines in the UK?" The answer is Yes. There is an Open Cycle Gas Turbine system operating in Spalding, Lincolnshire.

Combined Cycle Gas Turbine Power Station

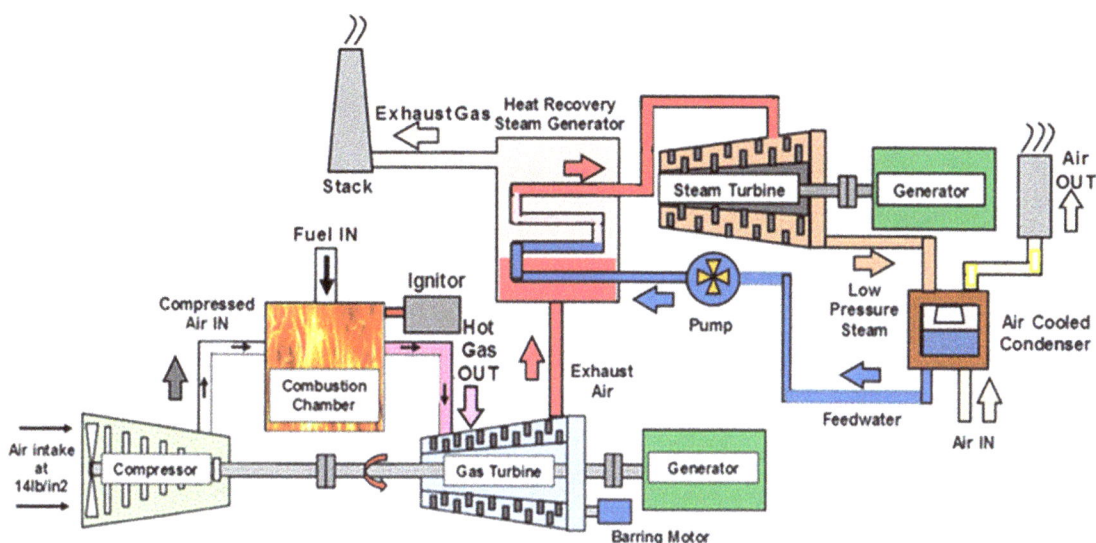

Fig. 58

The Combined Cycle Gas Turbine (CCGT) is shown in Fig. 58 above, and these are a little more complex than the Open Gas Turbine Drive. There is an added heat exchanger called a 'Heat Recovery Steam Generator' and two turbines, one gas and one steam, driving two generators. There are around 30+ Combined Cycle Gas Turbine Power Stations operating in the UK.

The Combustion Chamber and Compressor is the same as the Open Cycle Gas Turbine, but instead of taking the hot exhaust air straight to the stack having spun the 'Gas' turbine, to be vented, it is fed into a heat recovery steam generator, turning feed water into HP steam to drive a second 'Steam' turbine, and in turn a second generator.

The HP Steam then enters a air-cooled condenser under constant cool air pressure, which condenses the water to the base to then be pumped back into the heat recovery steam generator and turned into steam by the hot air from the gas turbine.

The Air-Cooled Condenser 'Air' is then vented to the atmosphere, having cooled the hot low-pressure steam into condensed water.

Efficiency is definitely up on the Combined Cycle Gas Turbine as against the Open Cycle Gas Turbine, because heat is recovered and produces steam to turn a second Turbine and Generator, but it will be much more expensive.

Both Gas Turbine Systems are **Non-Renewable** as the fuel is usually Natural Gas or Methane, although it is reported that Syngas can be used, which is a mixture of Hydrogen and Carbon Monoxide.

If Hydrogen could be used as a fuel in the Combustion Chamber, then this would become a **Renewable** energy.

The Combined Cycle Gas Turbine (CCGT) is not the same as the Closed Cycle Gas Turbine. They are two completely different systems. In a Closed Gas System, the exhaust gas is constantly being recycled through the Compressor from the Turbine.

Older Wind Power:

Fig. 59

Fig. 60

The Old-Fashioned Wind Machine concept of using wind to drive machines is by no means new. All that has happened is they have become more sophisticated and now generate electricity as the main objective.

Once the machine was built the energy was free and would operate 24/7. There is not a Western Film on the television that does not feature a wind machine similar to Fig. 59 above, and Norfolk is strewn with the old-fashioned windmills as in Fig. 60.

The Mid-Western Wind Machine's objective, in Fig. 59, was to mainly pump water out of the ground for cattle, or anywhere it was required. John Burnham and Daniel Halladay were responsible in the mid1800s for the development of this machine in the mid-West.

The old-fashioned UK Windmill's objective in Fig. 60, is primarily to make flour from grain by converting the power of the wind into mechanical energy. There are many different styles scattered all over the country.

The Windmill 'Fantail' is a small windmill with smaller sails at the back of the 'Cap' which holds the main sails. This Fantail turns the Cap into the wind. These are mounted at right angles to the main sails so one or the other catches the full force of the wind.

I am sure that you can see many parallels to the modern Wind Turbine. The angle of the flaps within the sails act like aeroplane wings and make the sails turn. Wind Power is the favoured power to take over most of the Power Stations of the future and of course is Renewable Energy.

Modern Wind Turbines:

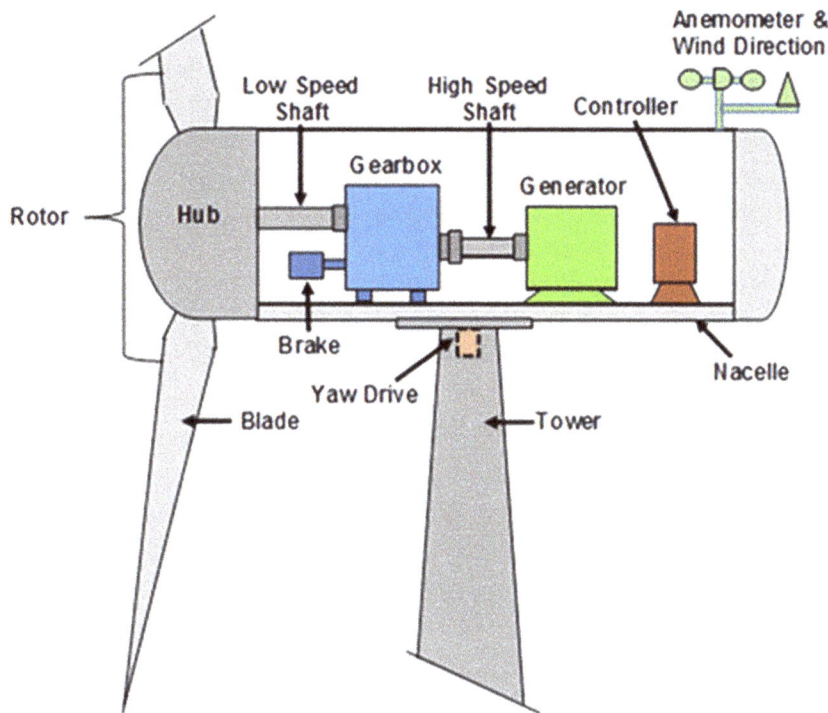

Fig. 61

The modern Wind Turbines' objective (Fig. 61) is to generate electricity, and they generate between 1 and 8 Megawatts on land and 7 to 12 Megawatts (MW) out at the sea. These Wind Turbines are usually in a cluster of around 10 – 12 and this is called a Wind Farm. In total the Wind Farm can usually generate over 1 Gigawatt (1000 Megawatts) of electricity. They are sited where there is a good prevailing wind, usually out in the open or high up. An ideal location is obviously out on the sea where there are no obstructions.

Can the wind be too strong? The answer is **"YES",** these turbines operate at a wind strength of around 10 to 50mph, any stronger is called the 'cut out' speed. If the wind is too strong, the Controller or On-board Processor will take measures to protect itself such as **'Furling',** where the Yaw Drive system turns the turbine away from the wind. The speed of the Turbines can be set to a fixed or variable speed.

Turbine Rotor Blade Pitch adjustment is an option if the wind strength is too strong. The 'Rotor' in Fig. 61 includes the 'Hub' and the 'Blades', which are shaped similar to an aeroplane wing, to obtain maximum turning power and are around 170 feet long. The weathervane gives information to the Controller, or On-board Processor and the blades can be hydraulically turned on the hub.

Depending upon the wind strength, which is measured by the Anemometer, the altering of the Blade **'Pitch'** is called **'Feathering'.** This feathering gathers more or reduces blade speed and is done automatically by hydraulics.

The Tower of course is hollow to allow Technicians and Engineers to climb up to the Nacelle at the top, which houses the Gearbox and the Generator. Also running down within the Tower are the HV cables, which if the Turbine kept turning (furling) round and round with the Yaw motor, they could twist badly and cause damage so there is an anti-twist mechanism.

The Gearbox is a very important part of the machine. Looking at Fig. 61, there is a low-speed shaft from the Hub to the Gearbox. This shaft is rotating at a speed of around **17 RPM.** This enters the gearbox and is increased to around 1500 RPM to drive the Generator.

There is a Brake on this shaft to prevent turning when maintenance is being carried out. There is another disc brake on the High-Speed Shaft local to the Generator. The Brakes are very similar to Disc Brakes used on a motor vehicle. The braking torque would have to be quite large to stop the rotor within a certain time. Being so large, an emergency stop would not be possible.

The Controller is the autonomous Processor of the Wind Turbine, and it controls everything from Blade Pitch and Rotor Speed, Yaw movement, voltage, and frequency of the electricity output. If the Controller considers the situation to be too extreme for the Turbine, it can shut it down. It is the Weathervane that has to monitor wind direction and speed to keep the turbine facing the wind, by sending information to the Controller.

Generator Type selection is very important as the wind will fluctuate. There are 4 types of Generators that can be used in Wind Turbines

1) DC Generator.

2) AC Synchronous Generator.

3) AC Asynchronous Generator. (Induction Generator)

4) Switched Reluctance Generator. We will discuss the different types in the book.

The largest Wind Farm in England is in the North Sea, off Hornsea. There are currently 174 Turbines in all, each turbine producing around 7-8 megawatts of power. The whole farm produces 1.2 Gigawatts. (12,000 Megawatts).

The Largest Wind Turbine in the world currently, is the French General Electric Haliade-X. It stands 853 foot high, about 5 times Nelson's Column in London. The Rotor Blades are 351 feet long, around the length of a football pitch, long and it generates around **12** Megawatts.

Cost always comes into the equation where these projects are concerned. The question has been asked **"why do they not build them all in the sea and not on land?"** The answer is probably cost and time. On land the cost of each turbine is around **£1.5m/MW** compared with around **£4.5m/MW** offshore.

When looking at the cost, it must be remembered the large cables, carrying the HV electricity, come from each wind turbine, be it on land or out in the sea. Substations and Hubs must be built to handle the power going onto the grid.

As mentioned earlier, the On-Board Controller, which is a bit like a Processor, controls things like direction, blade pitch etc. from information from the Weathervane. This Processor can also be programmed to carry out direction, remotely.

It does not matter if the remote control is in another country, it can, for instance, carry out **'Feathering'** which is altering the pitch of the Blades or **'Furling'** which is turning the Turbine out of the wind. Or in very extreme weather, shutting down the Turbine completely.

As mentioned, Wind Energy is Renewable energy.

Sea Wind Turbines:

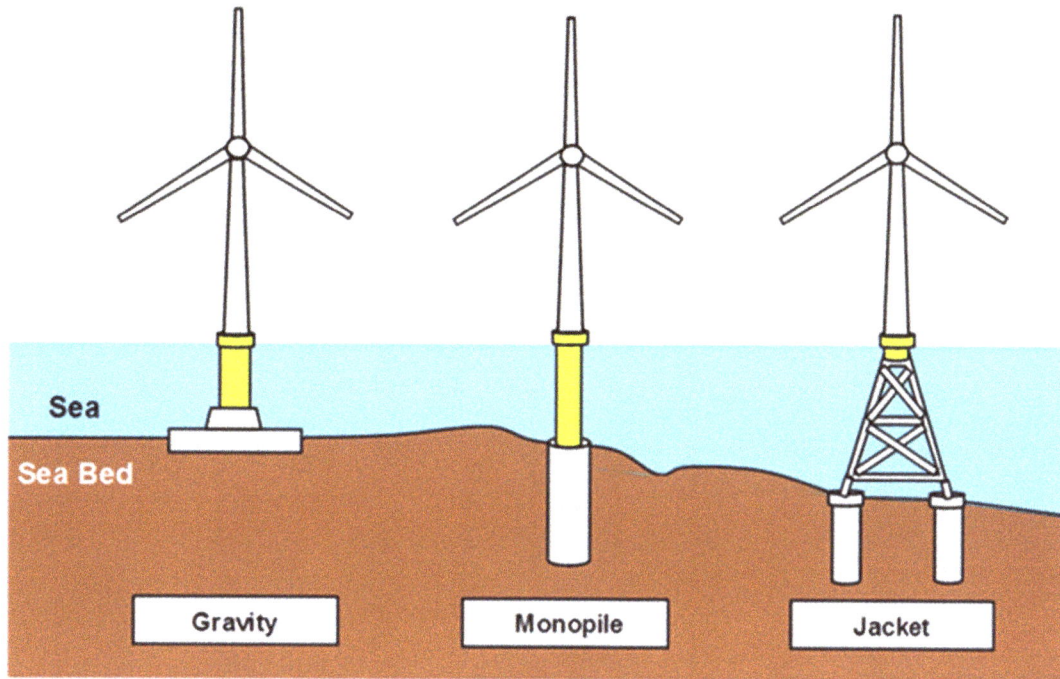

Fig. 62

Fixed Sea Wind Turbines cannot just be fixed anywhere in the sea, they can only be mounted to a depth of between **30 - 60 Metres** depending upon the type of design. Beyond this the option is 'Floating' Wind Turbines which is discussed next.

There are several types of Fixed Sea Wind Turbines, three designs are shown in Fig. 62, and the depths they can be installed are as follows:

a) The **Gravity** Foundation Design can be installed where the sea depth is around
30 Metres. (98.5 Feet)

b) The **Mono-pile** Foundation Design can be installed where the sea depth is around
40 Metres. (131.2 Feet)

c) The **Jacket** Foundation Design can be installed where the sea depth is around
50-60 Metres. (196.9 Feet)

Building the Turbines out in the sea has its own unique problems. Let us look at the Mono-pile Foundation Design. This design is a hollow steel pipe around 16 feet wide for a 5MW Turbine and 33 feet wide for a 10MW Turbine and is piled hydraulically into the seabed to around 180 feet.

Time Factor is another part of the equation as to whether to build out at sea or on land. For example, if they planned to build 100 Wind Turbines, to build them in the sea would take maybe around 8 years, much longer than around 6 years to build them on land. This would mean that Government Date Targets on clean energy might not be met.

The Hub is another name for a Sea Substation. Large sea wind farms will have a **'Hub'** that may be fixed to the Seabed, just like the sea Wind Turbines. Remember for every Sea Wind Turbine there is a High Voltage Cable that requires connection to the grid via a Circuit Breaker.

Sea Wind Turbines are very vulnerable to l Storms and Tsunamis. Building all wind farms in the sea could also make us vulnerable to our enemies. Destroying the Sea Turbines would cut down our electric stability. Several present Wind Farms are built far out in the North Sea, to make the best of the strength of wind. Some are just a few miles from shore and can easily be seen from land, but just to demonstrate how far out some of these wind farms are:

1) East Anglia Wind Farm is located 43 KM **(26.7 Miles)** off the UK coast.

2) The Hornsea 1 Wind Farm is located 120 KM **(74.5 Miles)** off the UK coast.

3) The Hornsea 2 Wind Farm is located 89KM **(55.3 Miles)** off the UK coast.

4) The Dogger Bank Wind Farm is located 130 KM **(80.7 Miles)** off the UK coast.

Just to give you an idea, the UK owned Sea Wind Farms currently total forty, and are located in Scotland, Wales and England. There are around six Farms under construction and many future Sea Wind Farms planned.

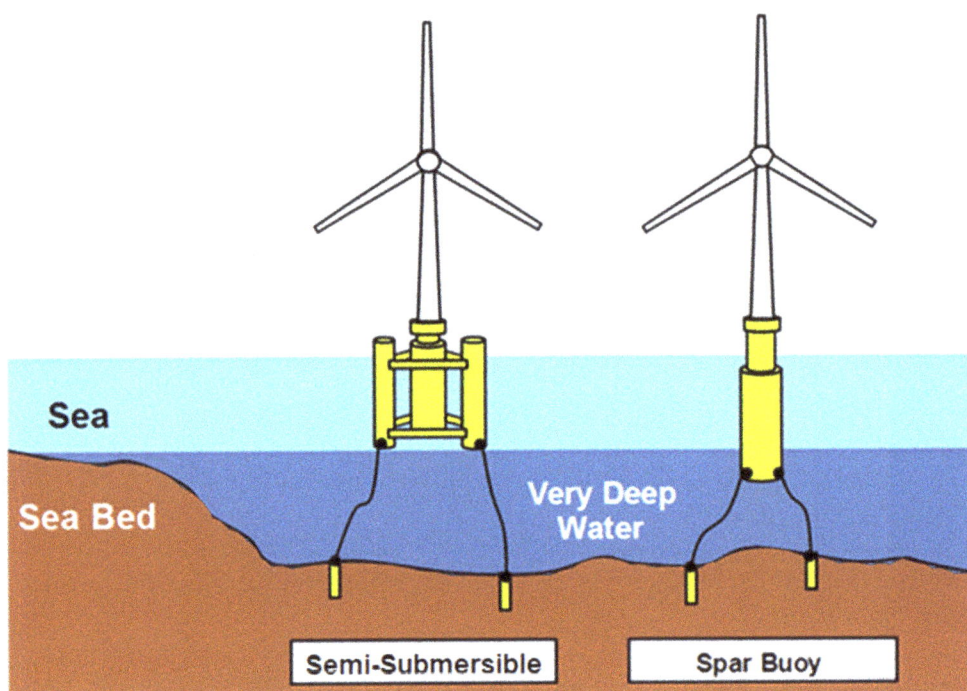

Fig. 63

Floating Sea Wind Turbines are totally different design. These can be installed by anchoring to the seabed (see fig.63, where the depth is deeper than **60 Metres.** The two types shown in Fig. 63 above, are Semi-Submersible and Spar Buoy. The maximum depth for a Floating Wind Turbine is around 300 Metres which is around **1000 feet.** Barge and Tension Leg are other designs.

Rough seas where there are very high waves. Floating Wind Turbines have to, as you can imagine, cater for many adverse situations out at sea. These turbines must have anchors fastened to the seabed which have to cater for tides and currents, so the depth may change between the float and the anchor. This means that the anchor cable must be long enough for high tides and to avoid fouling up at low tide.

Having the turbines so far offshore means there will be no fear of grounding on the seabed, so it's just the depth to cater for. The anchor and cable must not let the turbine move from its rest position too much. Remember that there are other Floating Turbines on the Farm and not too far away.

Anchors: There are three types of Anchors that can be used on the sea bed 1) The Drag Anchor, weighing around 50 Tons embedded into the seabed. 2) The Pile Anchor piled into the seabed. 3) Suction Pile, where the sealed top of a hollow tube creates a suction. Anchoring these Turbines presents all sorts of problems including Aerodynamics of the Turbine itself, with its blades both stationary and revolving, not to mention working at depth to install the seabed anchors.

High Voltage Cable(s) come from **EACH** of the Wind Turbines out in the sea. These run power from the Generator in the Nacelle. The electrical power in these cables make their way into the UK National Grid. The HV Cables that come from the Floating Wind Turbines must also be protected as far as possible in rough seas, as well as protecting them from things like surge tides and storm winds that happen in certain atmospheric circumstances.

Shipping has also to be catered for. It is not so bad in good weather where the ships can physically see the Wind Farms, as well as have them located on their charts and navigation systems, but what about fog? Many people make statements like, "Oh build all of the Wind Farms out in the sea" without looking at all of the problems.

The cost of Floating Sea Wind Turbines is far above a Fixed Sea Wind Turbine at around **£8.8m/MW.** There is an actual Floating Wind Turbine Farm at Hywind, off the coast of Peterhead, in Scotland. The farm consists of six 6MW Turbines. Total cost around £264m.

The Hornsea Wind Farm off the coast of East Yorkshire has around 165 Wind Turbines and the Farm is situated around 50 miles out to sea. It would be very impractical and very messy to run that amount of HV cables to land.

All of the Wind Turbines in the Sea Wind Farm take their individual cables to a Substation out in the sea called a 'Hub' and only a few cables come ashore. These are called **'Export'** cables. I have mentioned earlier, building Wind Farms so far out in the sea does make us very vulnerable to our enemies, so we must ensure that land-based Power Stations are kept as back up.

Vertical Axis Wind Powered Generators:

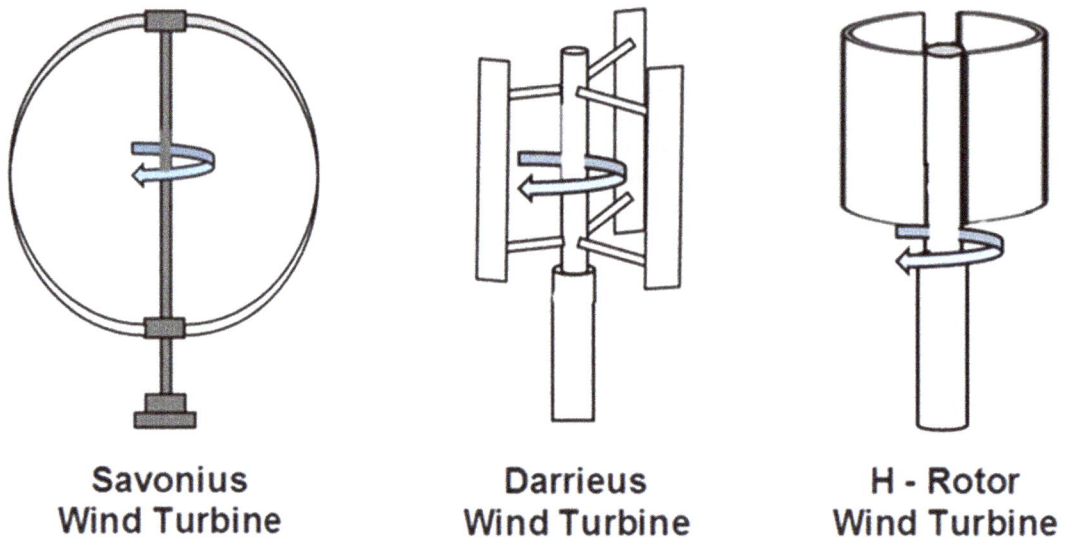

**Savonius
Wind Turbine**

**Darrieus
Wind Turbine**

**H - Rotor
Wind Turbine**

Fig. 64

Vertical Axis Wind Turbines are quite common. They have their rotating shaft, as the title describes, on a vertical axis as opposed to the very large turbines which are horizontal. Three designs are shown in Fig. 64 above, but there are many more different designs each one having a specific name.

Wind Turbines for domestic use would usually have to be the Vertical Type as they take up less room than a Horizontal Type. Planning Permission may be required, which is up to the local authority. Remember the Vertical Turbine must be mounted where it will catch the most wind, to be efficient and are usually mounted on the end of a very high pole.

The Output of the Generator would usually be around 2-3KW to give a good energy saving investment and remember unlike solar cells, which require daylight, these Wind Units strategically placed could give electrical energy 24-7.

For Domestic use, to run the whole house you are looking at a turbine what is 9-10KW. Remember it does not stop with the Turbine on a pole, there are the power cables to run from the Turbine to some sort of distribution system in the building.

The Generator is located somewhere below the spinning shaft on the same axis, and for domestic running of the property, it will be physically quite large. It is said that Vertical Axis Wind Turbines are more efficient than their counterpart Horizontal Turbines and may in fact be the Wind Turbine design of the future.

"Can the Wind be too strong?" Is a question commonly asked by potential customers before purchasing one. With a Horizontal Wind Turbine, it is possible for the weather to be too windy, and the On-board Processor called a **'Controller'** has to complete a task called **'Furling'** (turning the turbine out of the wind) or **'Feathering'** (altering the pitch of the blades) to compensate.

Vertical Axis Turbines do not have this trouble and can cope more with strong winds, although this must be taken with advice from the supplier. These Turbines will not require a Controller as such, although they may have a degree of control i.e. being able to stop it

Wind Powered Generators (No Blades):

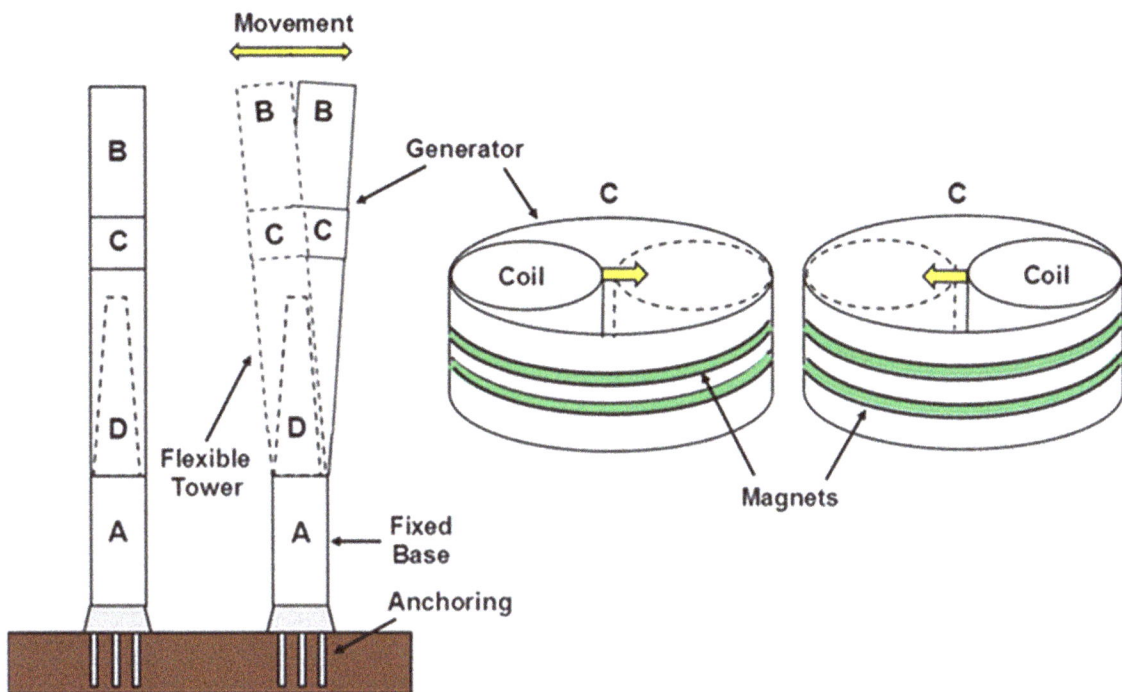

Fig. 65

Bladeless Wind Turbines are a breakthrough. If, a few years ago, you said that there are Wind Generators with no blades people would not have believed it, but the construction of the design of the Wind Generator that we are going to discuss now, strangely enough, has no blades.

First a vertical tower with a fixed **base** is installed which stretches up the tower to roughly one quarter of the height, as in the Fig. 65 above. The fixed base is anchored deep into the ground, and it will become obvious why as we go on.

The Main Tower 'B' is then installed above base 'A' with the generator 'C' mounted around three quarters of the way up the tower. The main tower 'B' is 'flexible' between it and base 'A' and will rock backwards and forwards in the wind.

Inner stop 'D' prevents the main tower 'B' from flexing too far. The tower will rock up against and stop and there will be a lot of force exerted on the base.

The Generator 'C' is mounted around three quarters up the tower and consists of a coil unit inside an outer magnet as in Fig. 65. As the tower rocks backwards and forwards in the wind, the coil unit moves from side to side inside of the magnetic field, which is created by the outer magnets thus causing a huge EMF which runs down the centre of the tower.

The advantages, if this was to be a viable proposition, would be that there are no gearboxes or moving shafts, no large Induction Generator etc. However, there would be quite a lot of movement of both the tower in the upper section and the coil units inside of the generator.

The problems, which I am sure that the designers have thought about, that I personally can see, is that the whole unit cannot be turned out of a very strong wind like the Horizontal Bladed Wind Turbine and so there will be a lot of strain both on the base and the flexible joint between the base and the tower

Older Water Power:

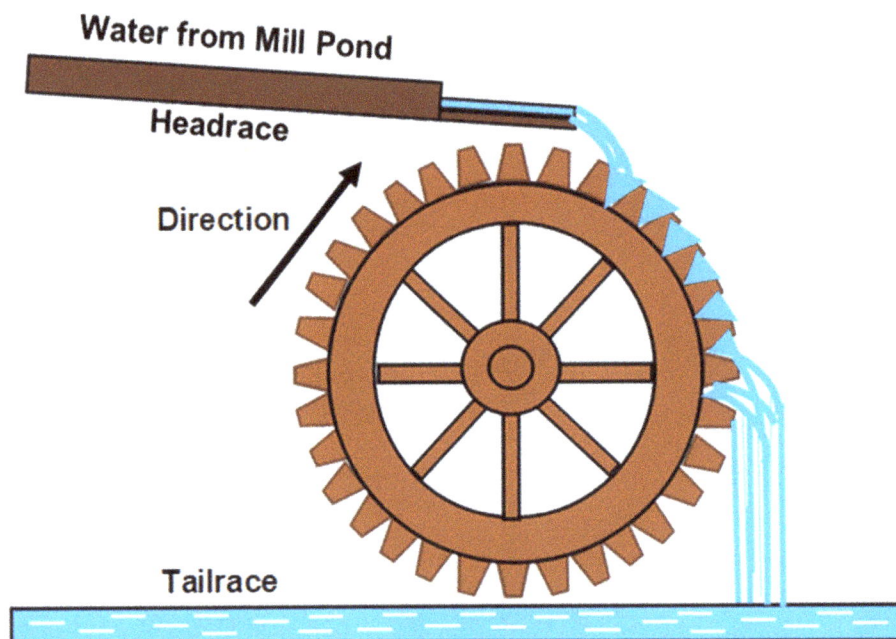

Water from Mill Pond

Headrace

Direction

Tailrace

Fig. 66

The Water Wheel, as in Fig. 66 above, has provided waterpower for some considerable time, although not for generating electricity. These Water Wheels are scattered throughout the country in all sorts of shapes and forms.

The Objectives of the Water Wheel was to transfer the power of the water to turn a large wheel, which turned shafts on machinery inside a mill making textiles or turning wheat into flour. The Feed Water is from a '**Mill Pond**' and this flowed down a '**Headrace**' which is a chute and equivalent to a '**Penstock**' in a modern Hydro-turbine.

The water flows onto the protrusions, sometimes shaped like cups or buckets, and forces the wheel to turn. After exiting the wheel, the water drops into another pond called the '**Tailrace**' and eventually into another river or stream.

When you look at a water wheel there are many similarities in the way they work to a modern Hydro-turbine as follows:

1) The '**Mill Pond**', as the saying states, is a very still pond of water formed by damming a stream which in a modern Hydro System would be equal to the '**Reservoir**' except not as large.

2) As mentioned, the '**Headrace**' is equivalent to the '**Penstock**'.

3) The **Water Wheel** of course would be the **Hydro-turbine.**

4) The '**Tailrace**' would be equivalent to the '**Outflow**'.

Look at the Water Turbine description and Fig. 67 in the next chapter to see the similarities mentioned above. The Industrial Revolution brought Water Wheels into their own. Once built the wheel would continue to function 24/7 as free power. John Smeaton and Benoît Fourneyron designed Water Wheels which were much more efficient. Waterpower is a Renewable energy.

Hydro Powered Generation:

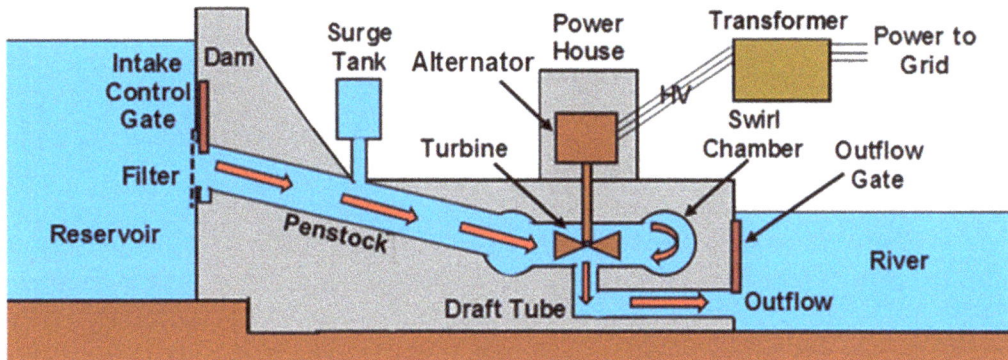

Fig. 67

Hydro Electricity is obtained by using kinetic energy in water to drive a turbine and in turn an Alternator. It produces around 10 Megawatts. Fig. 67 above shows the different parts of a typical Hydro Power Station construction.

The water leaves the reservoir through a filter, which stops debris from going down into the Turbine, and down to the **'Penstock'** to the Turbine Unit. This turns the Turbine and hence the Generator.

The 'Swirl Chamber', which has many different names, is next. This chamber allows the water to pass straight through the turbine via the **'Wicker Gate'**. Depending upon the turbine design, generally the water is forced down onto the Turbine Rotor. The water continues along the **'Draft Tube'** and via the **'Outflow'** into the river.

The Water Speed is controlled by the **'Intake Control Gate'** by opening the gate wider when the load is larger and closing it more if the load is lighter, and it can be shut off completely by closing the gate and the **'Outflow'** gate to allow for, say, maintenance. The 'Surge Tank' dampens any surge of high pressure and is also a water store.

The Alternator is not much different from the Generators and Alternators, covered in earlier chapters, the only difference is that this set up is driven by a Water Turbine instead of a Steam or Diesel one. There are several systems that may differ from the one described here.

The Alternator in Fig. 67, is one method of using water to produce electricity and is called the **'Impoundment'** method and it is the most common method. The height the water in the reservoir, sometimes called the **'Head'** or **'Head-pond'**, and the angle of the **'Penstock',** all contribute to the Kinetic Water Energy available at the turbine.

One of the River Hydro Power Station types are called Impulse Stations, and are usually supplied via a river fed from, say, a lake instead of a reservoir. These do not require a flooded valley. A 'Low Head' Hydro Power System is where the head is less than 20 metres above the turbine.

This may be an excellent choice, as there is no need to flood a valley for a reservoir to provide the water feed, but estimates would have to be calculated as to how much kinetic energy there is in the flow rate of the water, and checks made to see if diverting the water would have any effect on industry, housing etc., downstream.

The Surge Tank in the diagram above is usually there if the Penstock is fairly long, and for fluctuating loads. If the load decreases the water backs up into the Surge Tank and if the load increases the water flows back out into the Penstock to supplement the loss.

The 'Pump Storage' Method is another type of Hydro System where water is pumped from a lower level, using other types of energy for the pumps, such as wind and solar, to a small reservoir at a higher-level supplying water to the turbine.

Turbine Methods are not the same for all types of Turbines:

a) **'Impulse'** Type Turbine - Low flow rate from rivers or streams.

b) **'Reaction'** Type Turbine - High flow from a reservoir through a Penstock as in Fig. 67 (see previous page).

Several 'Types' of Turbine of turbine are below. Each **'Type'** being a different type and shape of impeller blades used in Hydro Power Stations. Below are some of them and if they are Reaction or Impulse:

1) Pelton Turbine   (Impulse)

2) Cross Flow   (Impulse)

3) Turgo   (Impulse)

4) Francis   (Reaction)

5) Propeller   (Reaction)

6) Kinetic   (Reaction)

7) Kaplan   (Reaction)

8) Straflo   (Reaction)

9) Tube   (Reaction)

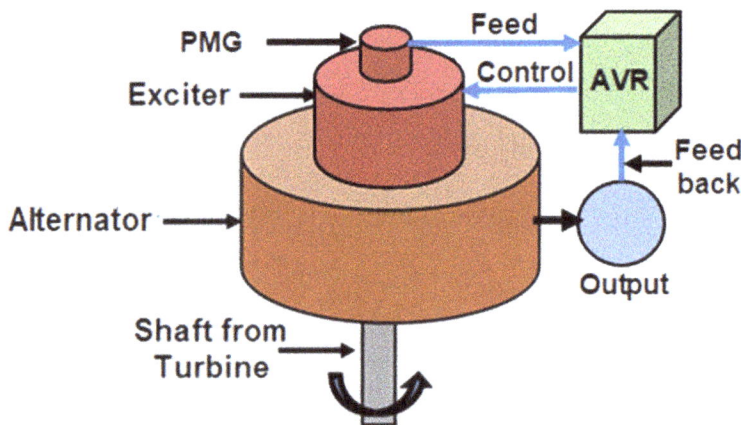

Fig. 68

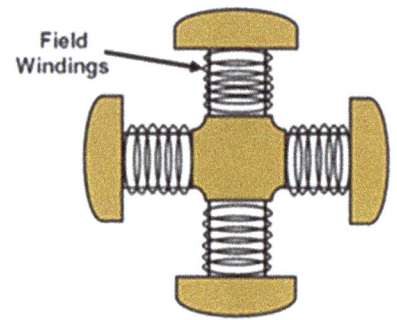

Salient Pole Rotor

Fig. 69

The Alternator in Fig. 68 above, is an example of one type of Alternator setup, this includes an exciter which is controlled via an AVR (**A**utomatic **V**oltage **R**egulator.) The exciter, if you recall in earlier chapters, supplies DC to the Rotor windings and here that could be a **'Salient Pole'** Rotor (Fig. 69). These are both driven by the shaft coming up from the turbine. Salient Pole Rotors are usually low speed around 1500 RPM.

The Salient Pole Rotor, Fig. 69, spins creating a Rotating Magnetic Field within a Stator, which is made up of Copper Wire, exactly like a Three Phase Motor, and it is from here that the HV Output Voltage is taken. The Alternator must be synchronised with the grid.

The Load Factor is equated from the maximum power that is available from the Hydro Power Plant running 24/7 against what is obtained from the plant. For instance, if the plant was running 24/7 and supplying a high amount of power onto the grid to assist in supplying cities, then the Load Factor would be very high. But if the Plant was only switched on in the evening to supplement power for lighting etc., then the Load Factor of the plant would be very low and inefficient.

The future of Hydro Power Station is good, as after the initial expense Hydro Power Stations just have maintenance costs with electricity 24/7. At present around 2% of the UK power is via hydro Power Stations.

A flooded valley would have to be created to form the Reservoir unless a **'Run of the River'** system is an option. This could disturb the eco system and displace people from their homes.

Apparently, Methane, caused by the flooding of vegetation which rots at the bottom of the reservoir, is given off in the Hydro Process. So, this process does not sound completely innocent.

Most of the Hydro Power Stations are in Scotland and Wales. I have been around the Hydro Station at Pitlochry in Scotland, which was part of the Tummel Valley Hydro Electric project. This station has a bypass for salmon to get to their breeding grounds upstream and is a sight worth seeing.

The number of Hydro Power Stations in the **UK** is around 100+, the largest being Cruachan Dam in Scotland, which produces 440MW. **England's** largest Hydro Power Station is Kielder Power Station in Northumbria, which produces 5.5MW.

In comparison, the largest Hydro Power Plant in the world is in China. The Three Gorges Dam is around 600 feet tall and 7600 feet long across the Yangtze River. Its Generators generate around 22,500 Megawatts of Electrical Power.

The main drawback of building a new Hydro Power Station, as I have mentioned, is that unless they build a 'Run of the River' Hydro Power Station, where they take the feed water off a river, then a valley would have to be flooded disturbing people and the eco system, to provide the waterpower.

Permanent: There is no doubt that the Hydro Power Stations that exist presently, are here to stay and now built, will provide 'Renewable' energy very cheaply well into the future, as the only cost would be maintenance and occasional renewal of the equipment in the system.

Having read this section, I will leave you to decide if this is the way to go for the future of Electricity Generation. Sometimes the most innocent of systems produce drawbacks and we are not going to be able to get rid of them all, although technology is constantly improving.

Wave Powered Generation:

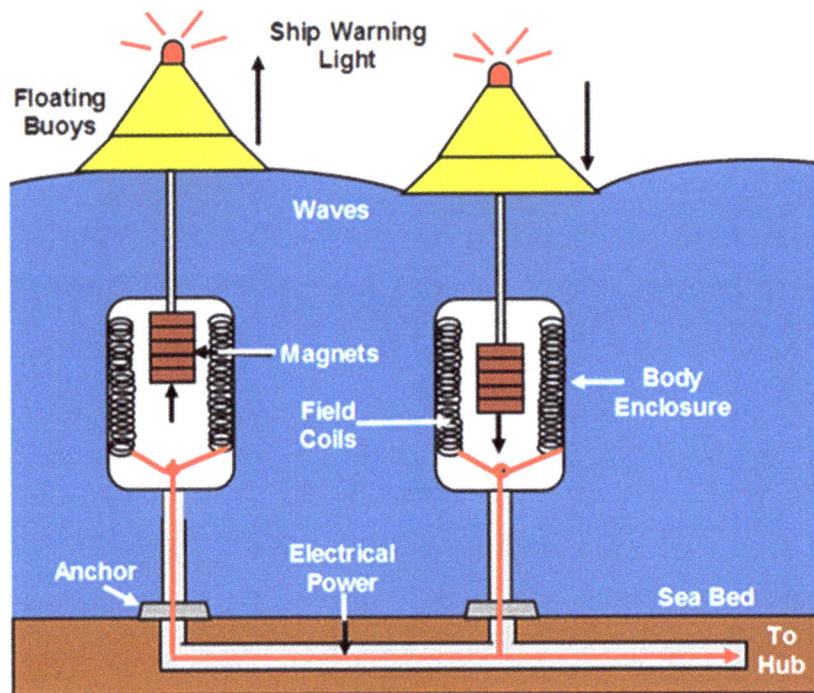

Fig. 70

Wave Generation Technology is something to look forward to for the future. Renewable Kinetic Wave Power is very environmentally friendly as no fuel is used to produce the power. Fig. 70 above, gives an example of one design of Wave Generation, but there are many other different designs some of which will be covered later. The advantages and disadvantages are listed at the end.

Linear Generation Buoys in Fig. 70, is one of those designs, where **'Buoys'** float on the sea surface and catch the waves rising and falling. Flashing Red Lights are placed on the top to warn ships/sea craft of their location. This type of system would be installed in shallow waters.

Fastened to the base of the buoy is a shaft which goes down into the top of the **'Body'** (grey) with magnets (red) fastened on the end of them, so as the Buoy rises and falls with the waves the magnets rise and fall inside of the body. The body is anchored to the seabed, so it does not move. See Fig. 70.

Surrounding the magnets are field coils, so as the magnets rise and fall, the magnetic flux cuts these field coils to produce electricity, which goes out of the Body via a conduit system and runs to a land 'Hub'

The main advantage of Wave Generation is that does not need fuel to run. The kinetic energy required comes naturally from waves. Generation is completed without any pollutants such as smoke etc., and after the initial outlay cost for the building, it does not require a lot of maintenance.

The main disadvantages are that it is a very challenging, costly installation. They have to be anchored to seabed to limit drift, as the sea is very unpredictable, one minute it can be calm as a mill pond and the next it can be stormy with high waves.

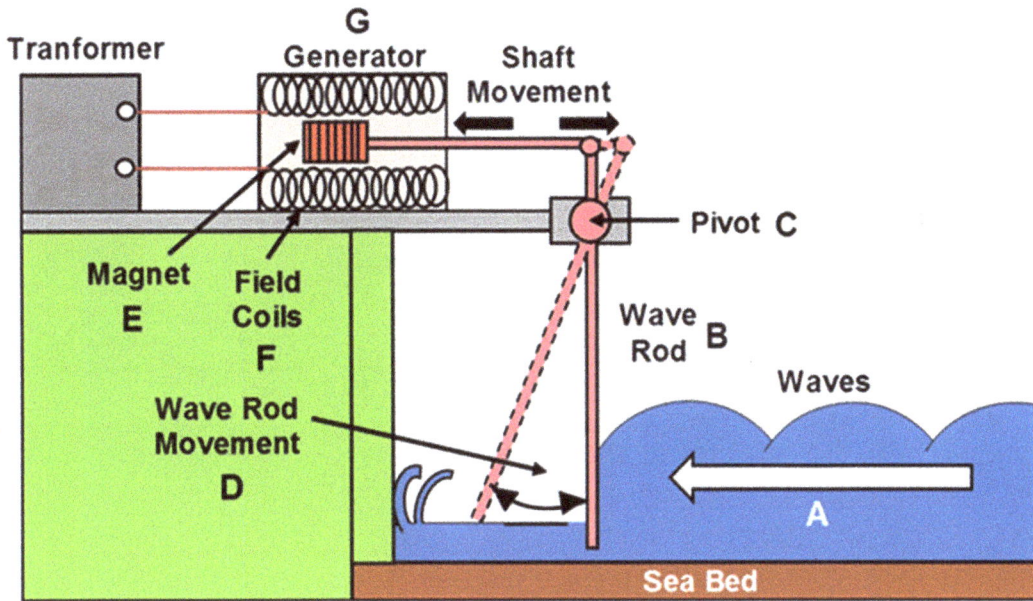

Fig. 71

Wave Rod System is one design where there is a 'Wave Rod' (B) which catches the kinetic energy (A) from each wave surge as shown in Fig. 71. This Rod moves backwards and forwards on pivot (C) with the wave movement and has a finite action (D). The wave rod (B) is connected to a shaft which causes the magnet (E) to move back and forth between field coils (F) producing electricity output to the Transformer.

This system must be constructed near to shore and can be very unpredictable. If the sea is calm, the Rod movement will be very limited but, in a storm, the movement can be extreme. Pivot (C) with its 24/7 movement is bound to incur wear. The Moving Magnet will be a Magnet and not an electromagnet.

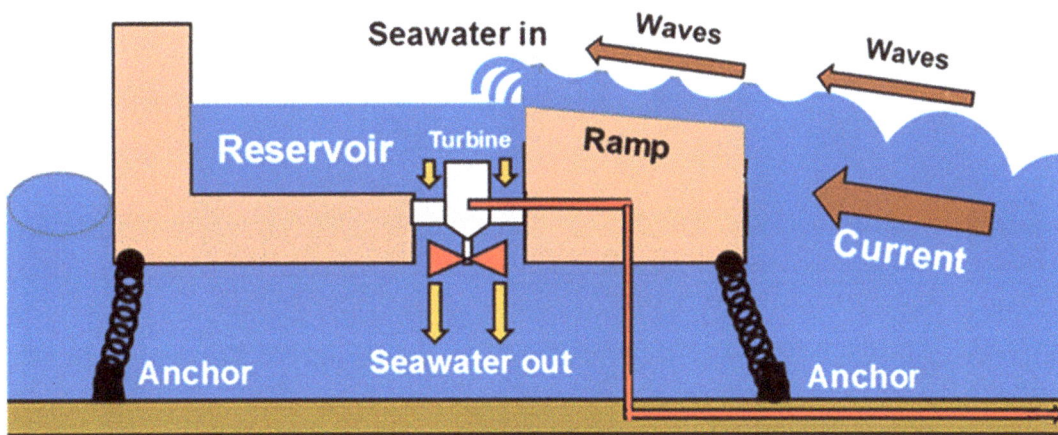

Fig. 72

Overtopping Wave Generation is where the force of the waves washes the seawater up the Ramp, as shown in Fig. 72 above, and into the reservoir. The Turbine Channel is the only way for gravity to restore the lower level and drive the water through, producing electricity which then goes down through HV cables in the seabed back to the grid. The problem here, is that without a lot of Current to create enough Kinetic Energy to force the water up the ramp, this system will be very sporadic.

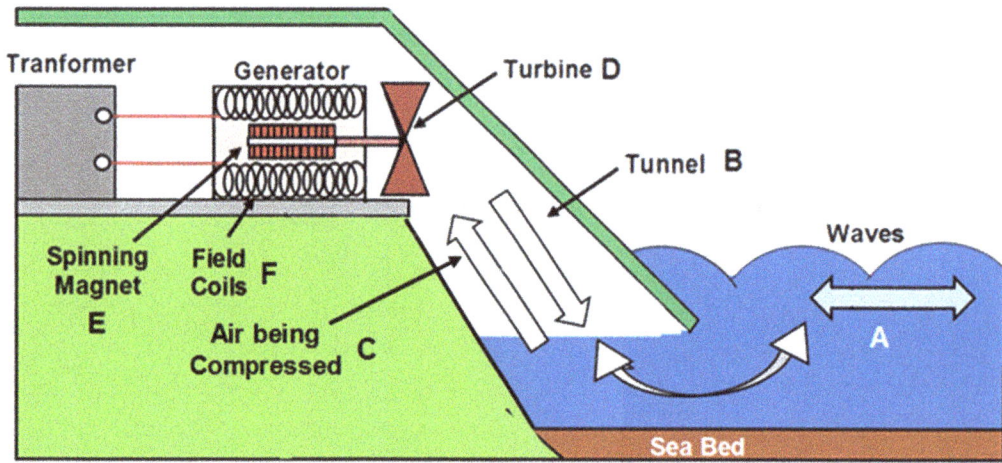

Fig. 73

The Oscillating Water Columns (OWC) in Fig. 73, shows the waves holding the kinetic energy **(A)** heading for the tunnel **(B)**. The wave force at the base of the tunnel acts like a piston compressing the air in the tunnel **(C)** and the compressed air then heads up the tunnel and turns a turbine **(D)**

By turning the turbine **(D)**, the shaft spins a magnet **(E)** which is connected on the end. Spinning the magnet **(E)** within the field coils **(F)** will produce an EMF which is connected to a Transformer.

With the Oscillating Compressed Air System, apart from the Generator, there are no moving parts, so once the actual system is built maintenance is only focused on the Generator. Obviously, this is a near to shore set up at the moment.

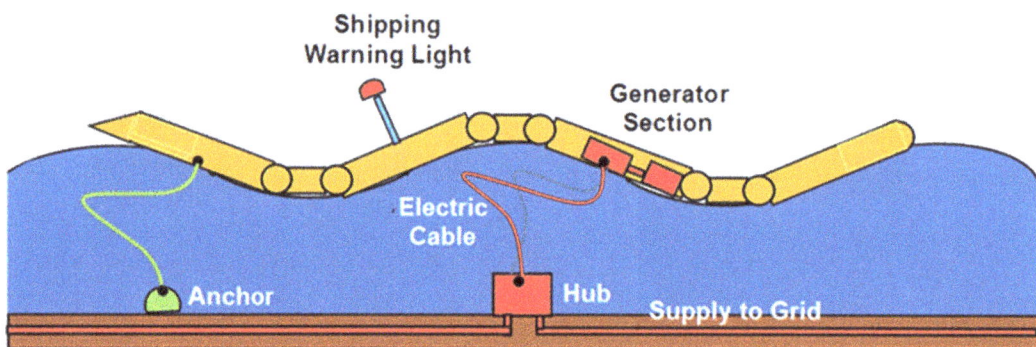

Fig. 74

Attenuator Hydraulic Wave Generation as in Fig. 74 is one design of Wave Generation where the waves holding the kinetic energy moves the floating sections up and down. All sections are connected with a system allowing them to pivot like a huge snake.

The whole thing must be anchored down onto the seabed so that it does not physically move too much from its position. The unit will turn as the tide direction changes to obtain the most efficient position for catching the waves.

The Generator gets its kinetic energy drive hydraulically, meaning that as the waves move the machine up and down, the pivoting systems turn the Generator Hydraulically. Again, like many wave machines this unit requires a wavy sea before it will work efficiently and without wind and waves it is not much use.

82

Tidal Generation:

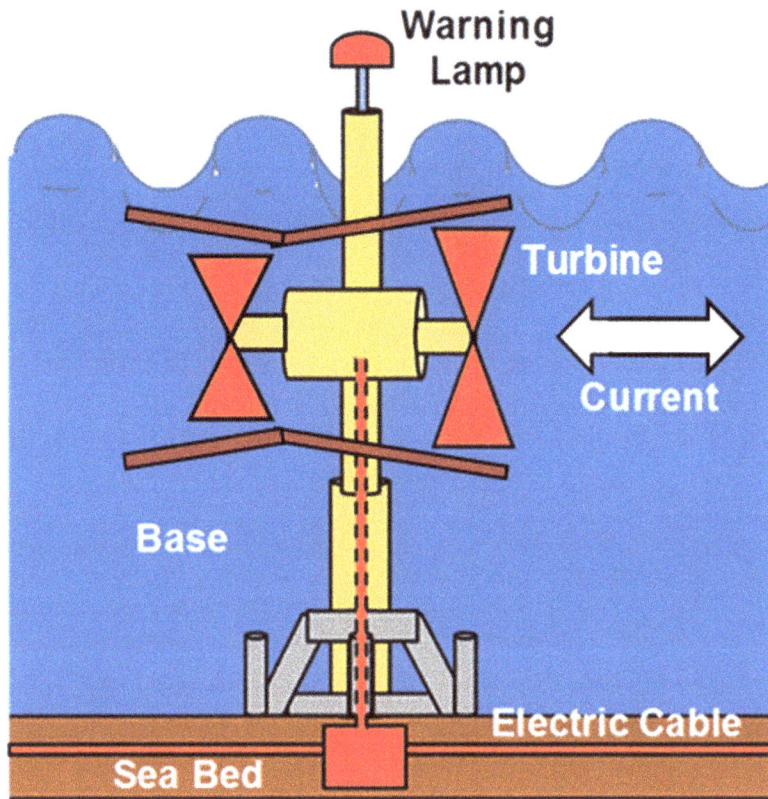

Fig. 75

The Fixed Tidal Generator System shown in Fig. 75, is one design of Tidal Generation, and like Wave Generation, has several designs. The initial design would have to consider several parameters such as, Cost, Shipping, Highest Tidal Current, Type of Tidal Turbine design, Electric Power Output etc.

The Turbine is usually located not too far from the shore, this means that it is not too great hazard for shipping, and it is not too far to run the Electric Cables. This is a simple design; the turbine is fixed to the seabed in the Prevailing Current Flow.

The Current is caught by the Turbine turning at a steady rate to try not to cause any risk to the marine life. The Fixed Turbine is designed for the blades to generate regardless of the direction of flow of the current, so the whole Turbine does not have to swing round like the Barrage Turbine.

One such occasion where a Fixed Turbine may be used is a dam across a river and in this situation, there may be several Fixed Tidal Turbines built into the dam. This could be as well as or instead of a Hydro Power Turbine.

There are pros and cons for all types of turbines. For instance, the cost of a current based Turbine which can be built anywhere, dropped into the sea and fastened to the seabed is fairly high combined with a limited output as in Fig 75, but much cheaper than the Barrage Turbine in Fig. 76 where a concrete structure has to be built on the seabed to house the Turbine.

Reliability: The Tidal and Wave Generation in terms of reliability is about equal to Wind Generation and in the future of Renewable energy could be its main competitor.

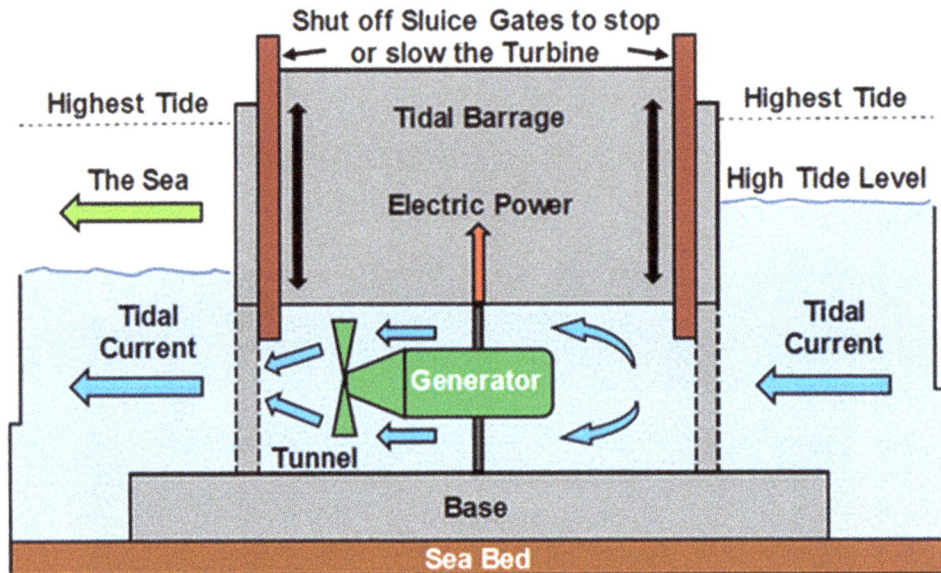

Fig. 76

Tidal Barrage Generation is another example of Tidal Generation, where a Concrete Barrage is built across or partly across, say the estuary, and many Turbine Generators such as the one in Fig. 76 are installed into it. The initial design would have to consider several parameters such as Cost, Shipping, Highest Tides, Type of Tidal Turbine Design, Electric Power Output etc.

The Turbine, unlike the previous 'Fixed' Type, can actually spin round to take into account direction of the Tidal Currents i.e. whether they are going out or coming in. This may not be so on every design. The idea is that, when the tide is going out then the water is high on the river side of the barrage, so the current will flow from the high side to the low side through a series of tunnels each of which will have a Turbine Generator installed so the Tidal Kinetic Energy will spin the Turbine Blades.

When the tide turns, the sluice gates will drop, or partly drop, trapping the water on the high side and when ready the gates will open and the current flow out towards the sea, i.e. in the opposite direction, and the Turbine will turn to capture the Kinetic Energy.

If the Electric Power is not required or some maintenance is to be carried out where they do not want the Generator to turn, then the Sluice Gates can close and stop the Turbines, so not so many Generators will be running.

Where is the largest Tidal Power Station? The Largest Tidal Power Station ever built is in South Korea and produces around 250MW of electricity.

Does Tidal Generation affect Marine Life? A barrage is a type of dam no matter how you look at it and dams must have an effect on the outflow of the tide. What about Marine Life that is a bit larger such as seals? I can see that there would be a problem to try and build a system that can cater for **EVERY** marine animal or fish.

The Cost will be higher building a Barrage Type Tidal Generation System because of the building of a structure i.e. the Barrage. This is of course Renewable Energy.

84

Ocean Thermal Energy Conversion (OTEC):

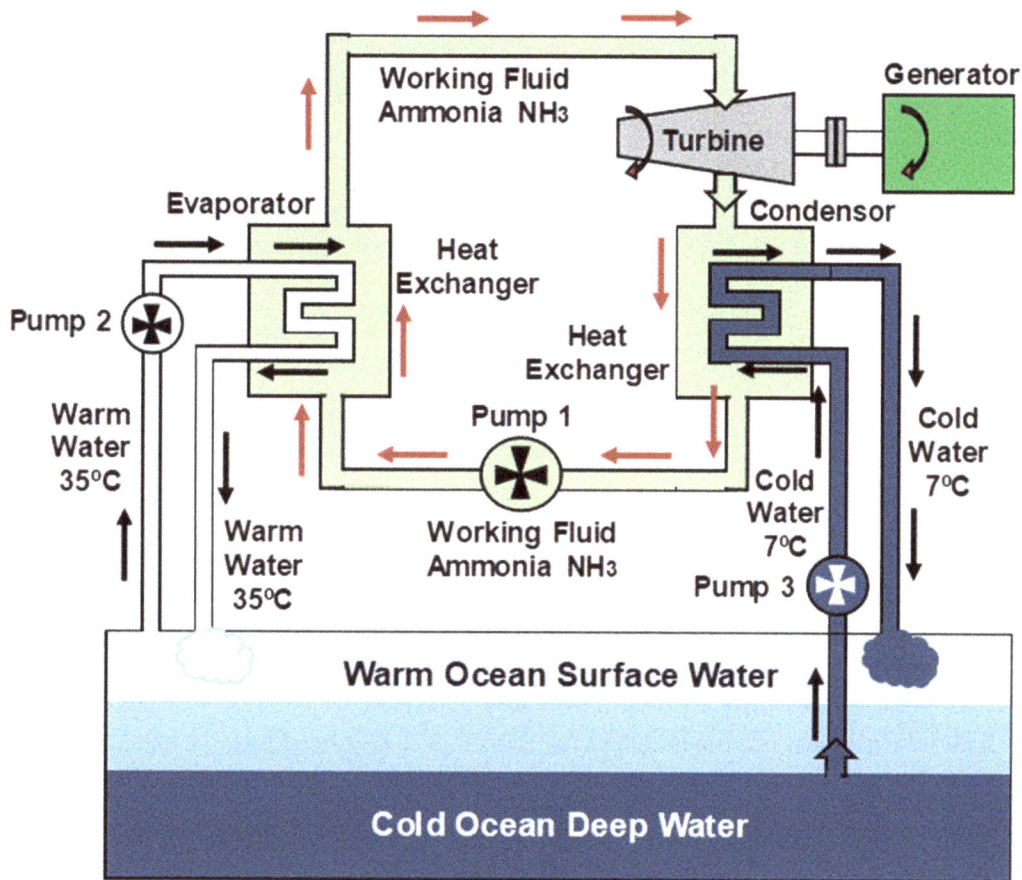

Fig. 77

The **Closed** Cycle Ocean Thermal Energy Conversion is quite a complicated system that can only be mounted in a tropical situation where the surface of the Ocean Water is relatively warm (35ºC) or above and the Deep Ocean Water, around 3000 Feet down, was much colder (7ºC).

There are two types of Thermal Energy Conversion system namely:

1) **The Closed Cycle System** sometimes called the 'Anderson System' which is shown in Fig. 77 above.

2) **The Open Cycle System** which is shown in Fig. 78 on the following page.

Firstly, let us have a look at the Closed Cycle (Anderson Cycle) OTEC. This system will certainly **NOT** be around the UK, as this system relies on the tropical ocean where the surface water is 35ºC or above.

The Evaporator and the Condenser which are two heat exchangers, a warm water one which is called an **'Evaporator'** and a cold water one which is called a **'Condenser'**. The Closed Loop Fluid in Fig. 77 is green and as you can see is in a Loop hence part of the name of the system.

The Working Fluid in this case is the Closed Loop Fluid, Ammonia. Remember that the warm fluid (ocean water) in the Evaporator is 35ºC or above and the cold fluid (ocean water again) from 3000 feet down is around 7ºC.

The Ammonia Gas from the Evaporator in the Closed Loop drives the Turbine and hence the Generator. After driving the Turbine, the gaseous Ammonia passes into the **Condenser** which is cooled by the 7ºC cold ocean water and at this point, the Ammonia Gas turns back to liquid and is pumped round, by Pump 1, in the direction of the red arrows to start the cycle all over again.

Just to sum up, when the Evaporator Heat Exchanger receives the Ammonia with its low boiling point, the warm water (35ºC) being pumped up by Pump 2 from the Ocean Surface and back down again into the ocean, causes it to vaporise and because the main Ammonia Pump is causing pressure in the system this Ammonia Vapour is directed into the Turbine.

The Three Pumps in the system are:

1) Pump 1 – the Main Ammonia Pump ensuring Liquid Ammonia is pumped from the base of the Condenser into the Evaporator.

2) Pump 2 – Warm Water Pump taking the surface water 35ºC from the ocean surface up to the Evaporator Heat Exchanger to heat the Ammonia.

3) Pump 3 – Cold Water Pump taking the cold water 7ºC from the deep down in the ocean up to the Condenser Heat Exchanger to cool the Ammonia.

Why is Ammonia, which is very unfriendly, used as the Closed Cycle Fluid? It must be a liquid with a very low boiling point and this in turn reduces the size of the Heat Exchangers and the Turbine. The Ammonia being in the closed loop will never be allowed to vent into the atmosphere.

The properties of Ammonia which is in the green Closed Loop of Fig. 77 are as follows:

1) Compound consisting of 4 Atoms. (1 Atom of Nitrogen & 3 Atoms of Hydrogen)

2) Not particularly friendly greenhouse gas but found in many things.

3) Colourless gas.

4) Flammable on its own – NO, but certain concentration mixed with air YES.

5) Smell: Very strong sharp nontolerant odour.

6) Can be called Hydrogen Nitride or Nitrogen Hydride.

7) Chemical Formula NH_3.

8) Melting Point: - 77.73ºC

9) Boiling Point: - 33.34ºC **(This is why it is ideal for the closed loop)**

Could this Closed Cycle System be installed in UK Waters? No, the UK Waters would be far too cold. The Open Cycle OTEC, explained next, would be able to be installed in waters that were 30ºC or above.

What type of Generator can be used? The Generator **that** could be used, for instance, could be an Induction Generator or a Brushless Alternator.

The system can be land based or floating out in the sea. The problem with floating systems is always dependent how rough is the sea, where it is going to be placed and how is it going to be anchored to the seabed.

Is this system Renewable Energy? Yes, as the only fluid besides Ammonia is water and the Ammonia is not released to atmosphere.

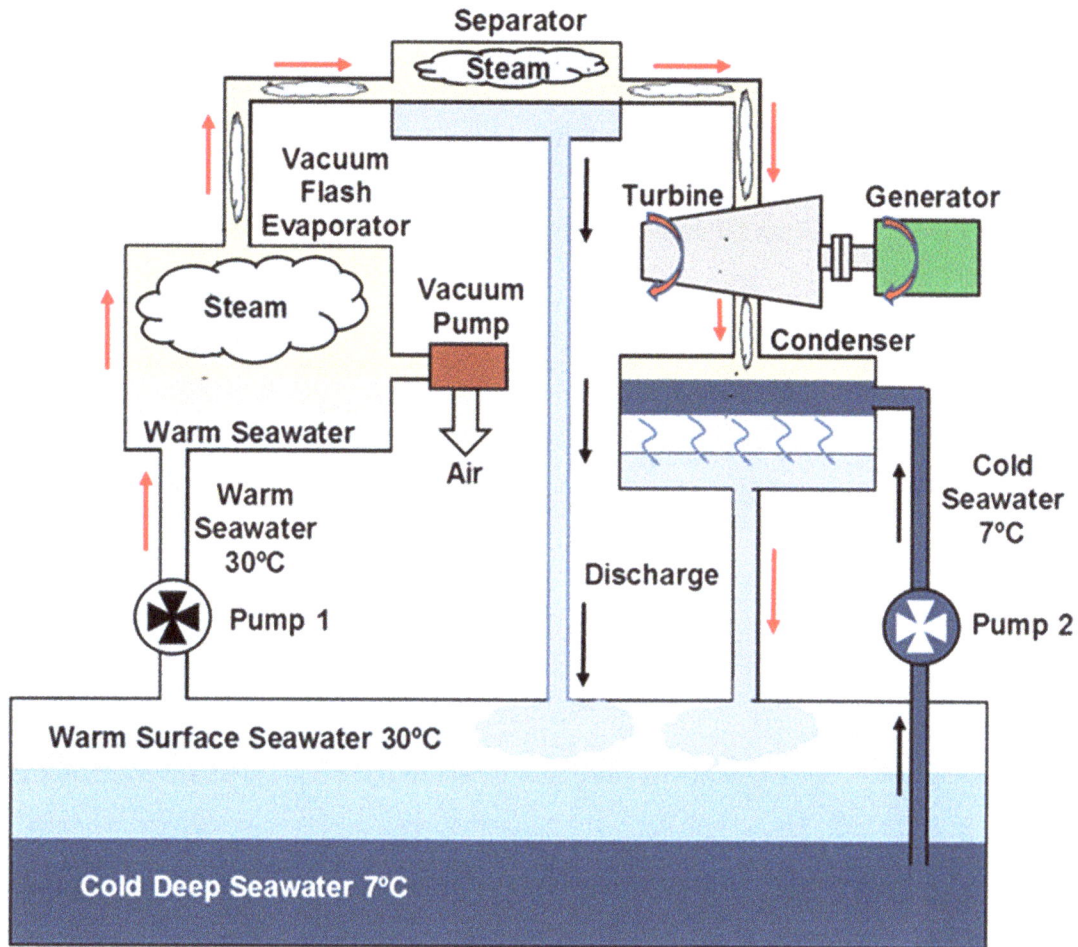

Fig. 78

The **Open** Cycle Thermal Energy Conversion in Fig. 78, unlike the Closed OTEC system which can only be set up in a tropical environment, can really be set up anywhere where the surface sea/ocean water is 30ºC or above. The good thing about this system is that it does not require a 'working fluid' such as Ammonia as the working fluid is in fact ocean water.

Land or Floating Systems are options, and Floating Systems can be built not too far away from the shore. Usually if it is a Floating System, it must be located in a position where it is sheltered from rough seas. The Floating System must be securely moored so that movement is very restricted, due to the Pipework and Electric Cables, so depth is limited.

The Vacuum Flash Evaporator plays a very important role here. This system relies on the fact that at atmospheric pressure, water boils at 100ºC, but take away that pressure and water will boil at a lower temperature.

Using Pump Number 1, the Vacuum Flash Evaporator has warm seawater (30ºC) pumped up into the Flash Evaporator (red arrows) and it has a Vacuum Pump taking all the air out so there is no pressure inside the vessel.

As the Sea/Ocean Water enters the vessel, there is no pressure, so it will immediately **'Flash off'** (Boil) and turn to steam. Salt will be left behind as the sea/ocean water enters the Vacuum Flash Evaporator. As the flashed steam will only be freshwater, in the case of the Open System this is called the Working Fluid as opposed to Ammonia which is used in the Closed Loop System.

The discarded water containing salt and other impurities is returned to the sea. The Separator is the next stop for the steam, where any excess water will be removed and again it flows back to the sea. The steam continues to turn a Turbine and hence a Generator.

The Turbine is turned by the steam which unlike Power Stations is at a very low pressure and temperature, but the Turbine is very efficient and so drives the Generator. After turning the Turbine, the steam passes into the Condenser where Pump Number 2 is pumping cold water at 7ºC from the sea/ocean depths to cool the steam and condense it back into water. This gathers at the base of the vessel ready for discharge back into the sea/ocean.

The Fresh Water in the Condenser will actually be fresh water which can be used for other freshwater projects rather than, as above, returning it straight back into the ocean where it would be wasted.

A Fresh Water Plant can also be constructed using this type of Open Cycle System, as instead of, or as well as, using this plant to purely produce electricity, with some slight modifications it can also be used to produce vast amounts of fresh water where there perhaps is a shortage. The Plants can be Sea/Ocean based i.e. floating or built on land with pipes running out into the Sea/Ocean. Separated salt would be a bi-product of the process.

Aas can be seen in Fig. 78, there is no Ammonia involved in the system. Instead, there is an open loop of water from the warm surface water to the cooler deep water making the system much safer. Hybrid Systems are available which combine Open and Closed Cycle Systems, but these are quite complex and expensive systems and still incorporate the unpopular Ammonia.

The Generator again can be an Induction Generator or Brushless Alternator. The load must be taken into consideration in this instance being a low-pressure system. I have included this Generator 'Drive' for interest.

Fig. 78 looks quite simple and gives an idea of what the system is about, but in practice this is not necessarily so, and it may be quite expensive and limited to set up. Nevertheless, we can class these OTEC systems as Renewable energy.

The world's oceans cover approximately 70% of the surface of the planet, so it is about time we started to put that to better use. These two systems, Open & Closed, Ocean Thermal Energy Conversion (OTEC) together with Tidal and Wave Power are doing just that.

Indirect involvement with the oceans includes Fixed Sea Wind Turbine Farms, Floating Wind Turbine Farms and Floating Solar Panel Farms. All these systems are also Renewable energy.

Geothermal Power Stations:

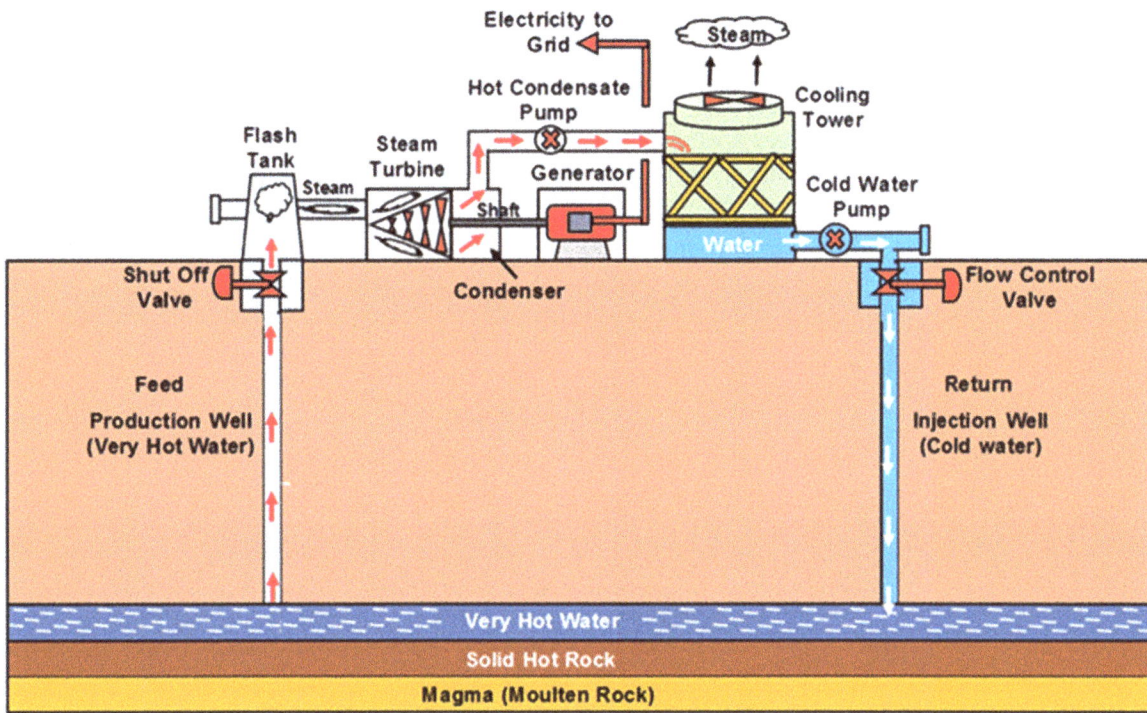

Fig. 79

Deep Geothermal Energy is obtained from deep down in the Earth. Shallow geothermal energy involves solar heat and will only provide heating warmth. Usually, these systems would be installed where there is Geothermal Energy or activity, which can be as much as a mile below the surface.

The system involves providing a 'Feed' or **'Production' Well** and a **'Return' Well** as in Fig. 79. It works when the valve opens, and water is injected down a 'Return' Well by the cold-water pump on the right of the diagram. This forms a water table above the hot rock deep in the Earth and is then superheated by geothermal activity.

Above the 'Production' Well is a Flash Tank to provide steam to drive a Turbine. Water returning from below ground will be around 180°C+, an extremely high temperature, and it will immediately turn to steam in the Flash Tank. The Steam is used to drive a Turbine and hence the Generator, where the generated electricity goes onto the grid.

After the Turbine, the steam enters a Condenser and now turns to hot Condensate which is then cooled by the Cooling Tower. After cooling and collecting in the cooling tower basin, the cold water is then injected via the cold-water pump down the **Return Well** into the Water Table and the process starts all over again.

There must be Control Valves in the system to control the process. An 'Emergency Shut-Off Valve' to operate in a Catastrophic Emergency or for, say, maintenance on the Turbine or Generator in Fig. 79. A Flow Control Valve to needed to ensure that the Cooling Tower Basin/Pond at the bottom does not empty.

The advantages are, that after the initial cost of the equipment and installing the wells, the Renewable energy is free, thanks to the geothermal layer. There is no CO_2, although with the very high temperatures involved, Hydrogen Sulphide H_2S emissions could be a possibility using this process.

The Disadvantages are that the wells are deep, and installation is very costly. These Wells cannot just be installed anywhere, they have to be located near or over a **'Geothermal Reservoir'** and in the UK, these are few and far between.

Grants have been issued by the Government for the processes to go ahead in the UK, but nothing has come of them. Shallow Geothermal Systems exist, but mainly just for heating rather than generating power. After Fracking had the very bad publicity, as allegedly, it was causing Earth Tremors and it involved drilling through the water table etc., it is unlikely that Geothermal Energy will be a UK venture at the moment.

This form of Generation is a Renewable energy, as it will not run out in the future, and a no greenhouse gases are released. However, as there are only certain sites where it can be used. Geothermal energy could be used, with the right safeguards, to heat homes as well as producing electricity.

There are three types of Geothermal Power Station:

1) **Flash Steam** – Possibly the most common as per Fig. 79, and the description above. And as mentioned it draws from a supply of water from below and turns that water to steam in a Flash Tank.

   Patents for this type of system have been applied for but there are no known plants in the UK.

2) **Dry Steam** – Rare Older Design. This type of system requires a large supply of underground dry steam where it is piped from the underground well, possibly miles down in the Earth, and directed directly into the Turbine.

There are no known plants either built or plans for one in the UK.

3) **Binary Cycle** – New Development. These use a type of 'working' fluid with a very low boiling point. Water at around 180ºC, which is not as hot as other systems, is pumped up into an evaporator where it flashes the sealed off 'working fluid' into vapour to drive a Turbine.

   After leaving the Evaporator the water is returned to the ground to be reheated and the working fluid after leaving the Turbine passes through a Condenser and the cycle starts again.

   There are no known plants either built or plans for one in the UK. If ever Geothermal Energy finally gets off the ground here in the UK, it will be the Binary Cycle design.

Iceland has this system of Generating Electricity where around 70% of their Energy is generated from nine Power Plants with 26% being Electrical energy. They have much more Subterranean Volcanic Activity than the UK. Building the power plant would be the expense the energy is mostly free.

Combustion Engine Generator Drive:

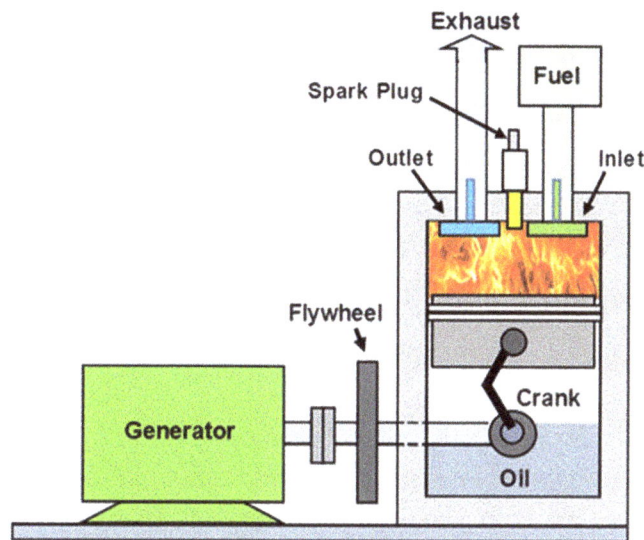

Fig. 80

Combustion Engine Generators are popular as both small domestic Generators and large Generation Units in the Chemical, Oil and Gas Industry. The Voltage Output and Frequency are fixed and cannot be adjusted manually. Car Generators come under this heading, although the output here will be very much lower in voltage.

Emergency Standby Generators are sometimes situated in places like a hospital, where several sets may be on standby in case of a loss of power. GPS (Generator Power Supply) are sometimes preferred as against UPS (Uninterruptable Power Supply) depending on the role required. UPS systems may be limited in current whereas the GPS Systems would not usually have this problem.

The Generator Engine Fuel is usually diesel for large sets and petrol for small Domestic type sets. They produce Carbon Monoxide (CO), so not too environmentally friendly. There are several types of diesels that are not too common. Biodiesel, which has animal fat and oils mixed with it, and of course **Renewable Energy**. Another is Emulsified Diesel where it is mixed with water is **Non-Renewable**. These Diesels have a finite life of around two years after which their Flash Points will be elevated and reach the stage where ignition becomes difficult.

Gas Driven Engines are available and are much cleaner than their counterpart liquid fuels. The three main gases used are Propane, Hydrogen and Natural Gas. These are less common than the liquid fuels, but Propane and Natural Gas are a **Non-Renewable** energy. It would be better for these generators to run on Hydrogen which is of course is **Renewable** Energy and far cleaner.

The Generator Size is in **KVA** because the output is both **'Active'** and **'Reactive'** power, which of course is in volt-amps (VA – Apparent power) not Watts (W – Actual power). VA does not take into consideration the power factor of the load (Cos φ)

Portable Generator Noise Levels depend upon the size of the machine and its soundproofing. Sixty – seventy Decibels is the norm. These must not be used indoors, due to the exhaust fumes of Carbon Monoxide (CO) which are present and require plenty of air circulation. **PUTTING THEM IN AN ENCLOSED BUILDING IS DANGEROUS!**

Turbo Expander Generator Drive:

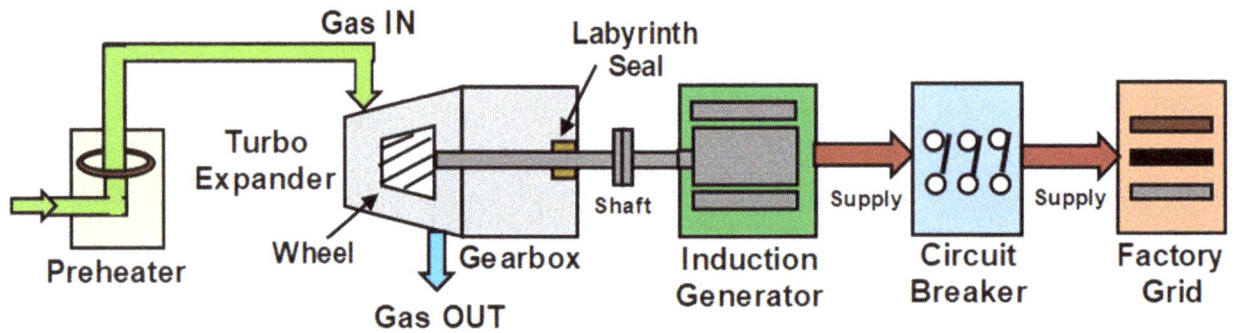

Fig. 81

Turbo Expander Drive is used quite a lot in the Chemical and Gas Industry. This machine (see Fig. 81) takes Kinetic Energy, trapped in process gases, through a turbine and turns it into Mechanical Energy driving, for example, as in Fig. 81, a Generator. The objective is to cause a cooling event with the gas.

The Gas Process undergoes various Isentropic processes, such as extraction of various gases, and can be used in conjunction with a **'Cold Box'.** High pressure gases are expanded to low pressure causing a cryogenic process.

Other names include:

1) Expansion Turbine.

2) Compander - may **be another name** if it is linked to a compressor.

The operation is carried out inside the Turbine by a cryogenic gas being forced onto a **'Wheel'** that catches the high-pressure gas, and as it passes through the Turbine causing it to turn as it expands. **The Joule-Thompson Effect** is what they call this process of cooling the gas in a Turbo Expander.

Induction Generators and Brushless Alternators are the type of Generator commonly used in Industry. Sometimes the Generator, such as an Induction Generator, is only used as a load, by means of using up the Kinetic Energy, and not necessarily used for producing large amounts of electricity, although it would still have to be linked to the factory grid.

Several Types of Turbo Expander are currently used in industry. This section focuses on the **'Cryogenic Refrigeration'** type used to lower the temperature of the gas. The end objective is to make the gas lose a large amount of its Kinetic Energy by making it do work i.e. drive a Generator or a Compressor. This will cause a Dampening Effect.

The Turbo-Expander Types include the following:

1) Direct coupling, where the turbine is directly coupled to the generator via a shaft and coupling. Remember the speed of the generator would be around 1500 RPM.

2) External gearbox between the turbine and the generator (Fig. 83). Speed has to drop from many thousand RPM to 1500 RMP.

3) Integral gearbox.

4) Multi-Stage can be used if there is more than one turbo expander.

Turbo Expanders by Guido Zerkowitz (Italian Engineer) in the 1930s. and first used in industry in the 1960s.

Conventional Bearings are sometimes used on the shafts, but magnetic bearings have been designed, which are much more efficient and reliable. These bearings do not require lubricant and have no friction losses.

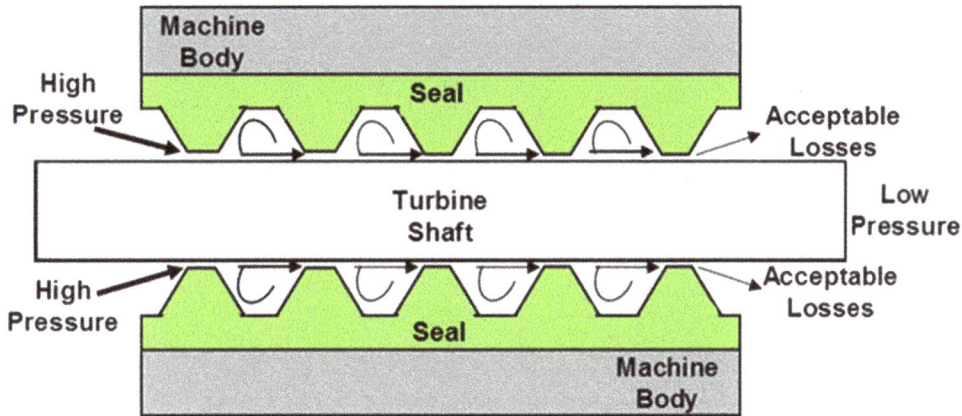

Fig. 82

Labyrinth Seals are usually used in a cryogenic situation as in Fig. 82 above. The diagram features a high-pressure end, forcing the gas to have an acceptable loss of seal between the actual Labyrinth and the shaft. Labyrinth meaning a system of complicated routes.

The Expander Speed can be around 30,000 RPM. If this is the case, then the speed would have to be dropped considerably to around 1500 RPM to drive the Generator. This would be achieved through a gearbox.

Exhaust Gases exiting the turbine have gone through the isentropic process which is less than -100ºC. At this temperature, depending upon the gas, it could turn to liquid.

Hydrogen Fuel Cell Generator:

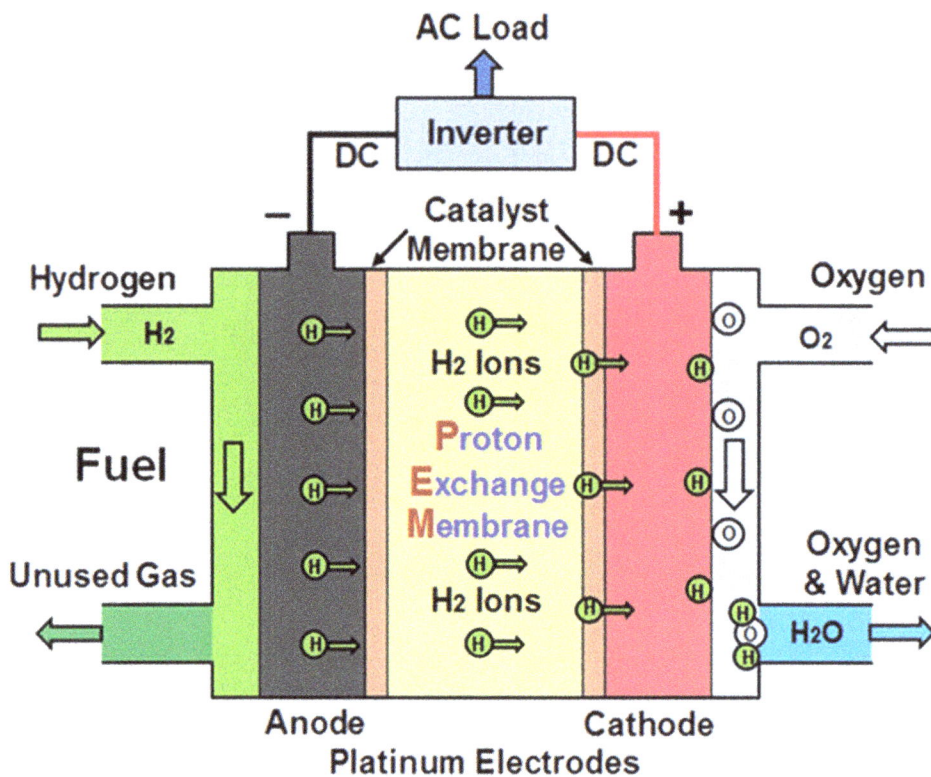

Fig. 83

The Chemical Generator shown in Fig. 83 is not an actual Mechanical Generator Machine. This Hydrogen Fuel Cell Generator is included here as it is called a Generator and is definitely a fuel of the future.

Note that this is a Fuel Cell **NOT** a Battery. Looking at Fig. 83 above, the Hydrogen Fuel Cell looks very similar to a battery, except that in this case the fuel used is Hydrogen and Oxygen, comes into the cell from the exterior and is not just a chemical reaction all contained within the cell. Also, fuel cells are constant and cannot go flat so do not require charging.

Hydrogen is the most abundant and simplest element in the universe. It is usually found combined with other elements. So, looking at Water (H_2O) we could split this formula into Hydrogen and Oxygen. That is why the main objective in exploring other planets is to find water, not just for drinking but for fuel and air. Some formulas that contain Hydrogen are as follows:

1) Water (H_2O). (2 Atoms of Hydrogen & 1 Atom of Oxygen)

2) Sulphuric Acid (H_2SO_4). (2 Atoms of Hydrogen, 1 Atom of Sulphur & 4 Atoms of Oxygen)

3) Hydrogen Chloride (HCl). (1 Atom of Hydrogen & 1 Atom of Chlorine)

The Fuel Cell Works by Hydrogen entering the Anode of the Cell. You will see from the diagram, that there is a catalyst (usually Platinum) separating the anode from the **Proton Exchange Membrane (PEM),** which in actual fact we could call the electrolyte. Some books call this the **'Polymer Electrolyte Membrane'.** Ions are formed from the Atoms after reacting with the catalyst (again usually Platinum) and move across the porous 'PEM' to the Cathode, where they are met with Oxygen Atoms. Excess Hydrogen is then recirculated.

Negatively Charged Electrons now flow out of the Cathode. The by-product from the Hydrogen and Oxygen atomic structures merging in the cathode is Water (Hydrogen & Oxygen - H_2O). The Catalyst is required to split the atoms, electrons and protons. The PEM does not conduct Electrons just ions and is made from 'Perfluorosulfonic Acid'

'Ions' are unbalanced atoms i.e. they have gained or lost their electrons giving them a positive or negative electrical charge. So, the number of protons (+) in the atom core do not equal the number of electrons (-) in their atomic structure.

The E.M.F. produced is DC, similar to a battery, and is then fed into an inverter to obtain AC to supply whatever load is required. Stacking the Cells to obtain higher power can be done, similar to Battery Units.

Types of Fuel Cell: There are five types are as follows:

a) Polymer Electrolyte Membrane (PEM) in Fig. 83.

b) Alkaline (AFC) – The main difference between this cell and the one that has been described, is that the Membrane/Electrolyte is an Alkaline Solution such as Potassium Hydroxide (KOH).

c) Phosphoric Acid (PAFC) – The main difference here is that the Membrane/Electrolyte is Phosphoric Acid.

d) Molten Carbonate (MCFC) – Uses Molten Carbonate Salt at around 600ºC as the Membrane/Electrolyte, so quite high temperatures involved.

e) Solid Oxide (SOFC) – Uses Solid Oxide material as the Membrane/Electrolyte, again around 600ºC.

Francis Thomas Bacon invented the Hydrogen Fuel Cell in 1932. In modern times Hydrogen is considered a Renewable energy.

Hydrogen Fuel:

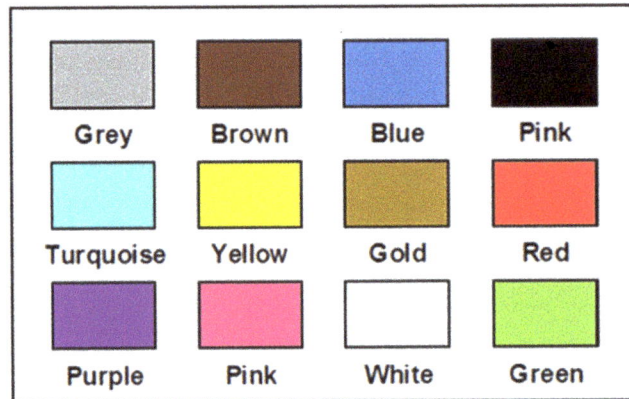

Fig. 84

Hydrogen, as I have mentioned many times, is the most abundant element in the universe and is a definitely the fuel of the future. You can be forgiven for thinking that there is only one Hydrogen type. It can however be found in **TWELVE** different colours (types) as Fig. 84 above.

Electrolysis is the method by which **Green Hydrogen** is achieved.

Hydrogen Types can be split into **Non-Renewable** and **Renewable** as follows:

Non-Renewable:

1) **Grey Hydrogen:** Made from natural gas (Methane)

2) **Brown Hydrogen:** Made from Coal.

3) **Blue Hydrogen:** Made from natural gas (Methane)

4) **Black Hydrogen:** Made from Coal.

5) **Turquoise Hydrogen:** Made using natural gas (Methane)

Renewable:

1) **Yellow Hydrogen:** Made from electrolysis using solar power.

2) **Gold Hydrogen:** Made by adding bacteria to spent drill holes.

3) **Red Hydrogen:** Made in a nuclear reactor. (Japan)

4) **Purple Hydrogen:** Also known as Pink Hydrogen. Made by extracting steam from nuclear power plants.

5) **Pink Hydrogen:** Also known as Purple Hydrogen. Made by extracting steam from nuclear power plants.

6) **White Hydrogen:** Produced through Serpentinization. (Water reacting with iron rich rocks)

7) **Green Hydrogen:** Generated from water by electrolysis.

I thoroughly believe that Hydrogen Powered Cars (**Renewable**), will become as common as Electric Cars in the future and will be cheaper than present Petrol or Diesel cars which of course use a **Non-Renewable. Energy source**

Ward Leonard Generator DC Speed Control:

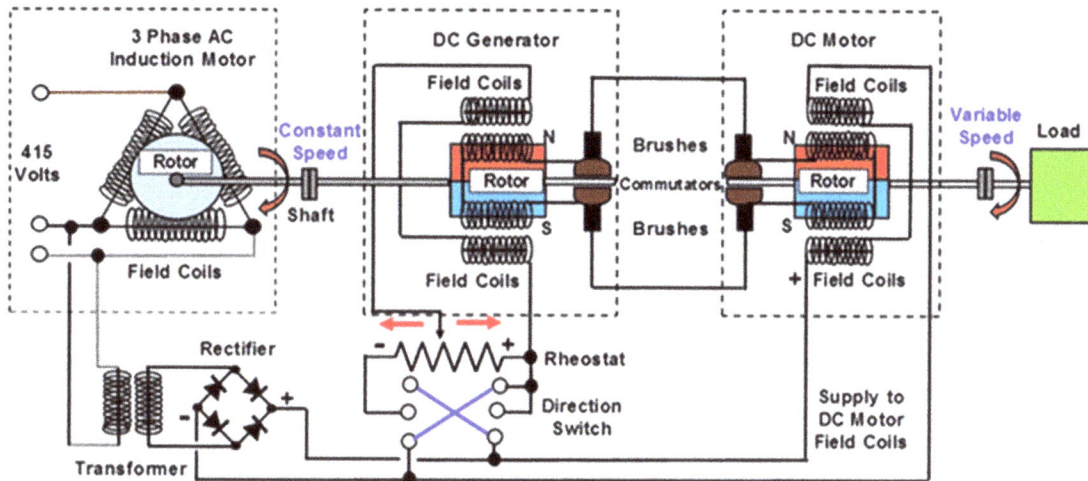

Fig. 85

The Ward Leonard System has been included for the simple fact that part of it is a Generator being driven by a Motor. Why is there an AC three phase Motor driving a DC Generator which feeds via its Brushes the Commutator of a DC Motor via the Motor's Brushes? The answer is Load Control. H. W. Leonard designed the system around the late 1800s. Applications are mainly Cranes & Elevators although other applications do exist.

The Constant Speed is obtained from a Three Phase Induction Motor, this can also be a Synchronous Motor or a Diesel Engine, which is connected via a coupling on the same shaft as a DC Generator. The Field Coil Strength of the DC Generator can be controlled by a Rheostat or Variable Resistance, which is shown with two red arrows, in Fig. 83, and it can be manually adjusted.

The Generator Output is via a Commutator and Brushes, which is fed to the Commutator and Brushes of a DC Motor. Now if the DC Generator turns, it will produce an output and turns the DC motor. The Varying Field of the Generator with the Rheostat can vary the speed of the DC Motor by increasing or decreasing the excitation of the DC Motor Rotor Coils. It is also a very efficient Brake.

The DC Motor Direction can be changed by changing the Positive and Negative Feed to the Generator Field Coils with a Direction Switch fed from a Full Wave Rectifier & Transformer. This of course would change the flux.

In the 1970s, on the top of two of our very large Acid Plants, around 180 foot up, we had a permanent Crane which moved from one side of the Plant to the other on two rails and was used on Major Shutdowns to lift and drop large tonnage vessels.

Being the 1970s, the Ward Leonard System was ideal as a control on these two Cranes. Full, smooth and immediate Speed Control and Braking was obtained, no matter what load was on the hook. AC Motors can have a Ward Leonard System, but a Rectifier or Diodes would have to be fitted.

The disadvantage of using this system would be cost, because of the amount of machines involved which also must be maintained, but the advantages outweigh the disadvantages! A modern-day crane control system uses Electronic Technology that will make the Ward Leonard System obsolete purely because of cost and the number of machines involved which all must be maintained. In its day the Ward Leonard System was the top system for Crane Control. Some of these systems may still be in use today.

Integrated Drive Generator (IDG) System:

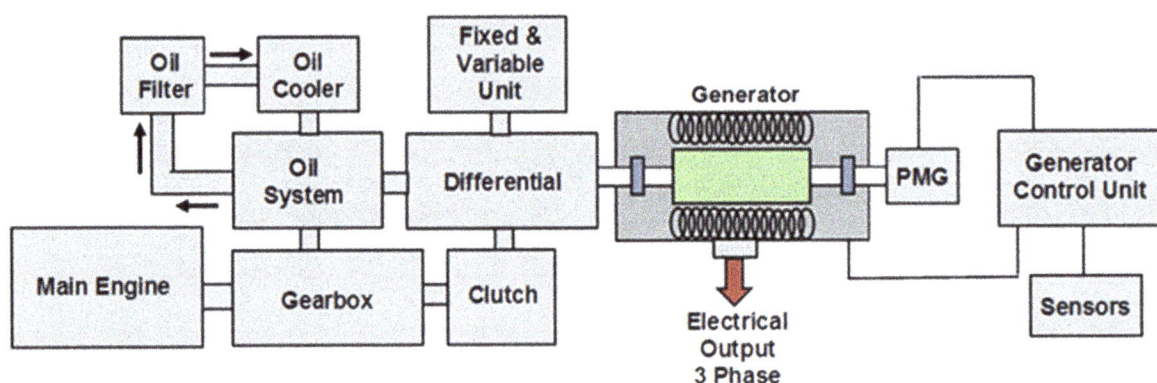

Fig. 86

The Integrated Drive Generator, as the name suggests, has the Drive and Generator Integrated into one unit. This Integrated Drive Generator (IDG) system is very complex, so I have given a very basic explanation and block diagram. Aircraft are the biggest users of the Integrated Drive Generators.

The Aircraft Electrical System needs to be powered with a constant, stable supply even if the engine speed may fluctuate, so is an extremely important piece of equipment. Flight Control and Navigation are two of the main systems that the IDC will power.

This is a Three Phase Electrical Power System. An AC Generator had to be found that could cope with all the stresses and strains of an aircraft, as it manoeuvres through the air. The supply originates via differential in the Gearbox off the Main Engine. The Frequency can be 400HZ.

Oil Levels must be monitored constantly to ensure efficiency. Low or High Oil Levels, in the sight glasses, can have a detrimental effect, as can contaminated oil. Engine and bearing wear rely on this commodity. Overfilling the Oil can do as much damage as a Low Oil Level. Oil is both a Lubricant and Coolant and the level is indicated on a Sight-glass. In most cases the Aircraft may have an oil Level and Pressure Sensor.

Oil also acts a Coolant as the temperatures can get quite high. After cooling the equipment, the oil flows through a filter into an external oil cooler where it is re-cooled before starting the cycle all over again. The oil can reach 180ºC+.

The Control System monitors all parameters such as speed, temperature, vibration etc. so any damage is limited.

The Generator Control Unit (GCU) controls the exciter field and hence the field of the generator. It is powered by the Permanent Magnet Generator (PMG). Sensor units provide the GCU with information such as engine speed, oil pressure, oil temperature etc. The GCU will disconnect the integrated drive if the oil temperature goes over a certain value, or if the oil pressure drops below a certain value.

Constant Speed Drive (CSD) controls the speed of the Generator and ensures a consistent Output Frequency, even under Engine Speed Fluctuation. The IDG Speed could be around 7,500 RMP gearing up to 12,000+ RPM. The Back-up System(s) are vital to such an important system as this, and an aircraft may have multiple IDC back-up redundant systems.

Shutdown Heat is the one thing to watch during maintenance. This equipment becomes very hot during running and can stay hot for some considerable time say, 1 hour!

Safety Wear (PPE) must be worn by Maintenance Personnel i.e.: the use of Goggles or Full-Face Mask for Eye Protection and Gloves in case of contamination and heat. Competency in Maintenance is a must being such an important system on the Aircraft.

Engine Starting must not be carried out when maintaining the IDG, and safeguards must be put in place to ensure this does not happen!

Hazardous Area: The only place that this is used is on an aircraft so it will certainly not be in **local** Hazardous Areas!

Solar Power Generators:

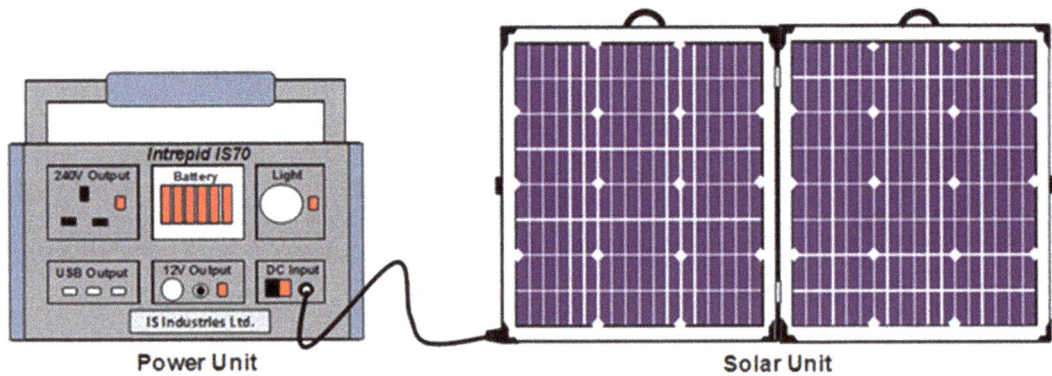

Power Unit   **Solar Unit**

Fig. 87

Solar Generators are not a Physical Mechanical Generator. This is not a magic box where we feed solar energy in one end and endless electricity comes out of the other. The set up shown in Fig. 87 above, is very portable, but some are not so portable as we will see. Cloudy conditions would not stop Solar Energy from working but not as efficient as a sunny day.

The Unit comes as two separate parts:

1) The Power Unit which contains Batteries, a Charge Controller and an Inverter.

2) The Solar Panel Unit which of course are the Solar Cells and folds up when not in use.

The Solar Panels and Power Unit Combined can now be obtained where the Power Unit has the Solar Panel built into the lid and top surface. The unit is around the size of a briefcase.

The Lithium Battery Power Output **in this set up** would be small in watt-hours (Wh) compared to a Machine Generator and very finite with its Electrical Energy. Nevertheless, it is very handy for a short camping trip or a day out. Lithium Batteries would take several years to degrade.

The Battery Charge is not solely charged by the Photovoltaic Effect off the Solar Panels, they can also be charged from the mains with a correct charger. However, the Power Unit contains a Charge Controller, so the charge of the Batteries is stable.

The Inverter is integral to enabling the change from DC Battery Power into AC power. One important factor of the output is the size of the Inverter, it needs to be large enough for the task. Check size of all specifications carefully to make sure it will meet your requirements, before you purchase.

Battery size is a consideration if they are to run a house. These Units can go up to 6KW which will go a long way to running most electrical items, but just remember they are Battery Units so the larger the total house KW demand the less time you will have on battery power.

The Inverter Size must also be taken into consideration and the amount of KW it can handle. You would have to be careful with the **'Diversity Factor'** i.e. what appliances are on at the same time. The larger the Unit the more the cost, and you may be looking at a cost of around £5000 for a 6KW Solar Unit with a 4KWh Battery.

The Weight will be quite high even for small units because of the Lithium Battery. The Main Unit could weigh around 60Kg (nearly 9.5 Stone) so the larger units would not be so light and portable. Solar energy is a Renewable energy.

Floating Solar Farms

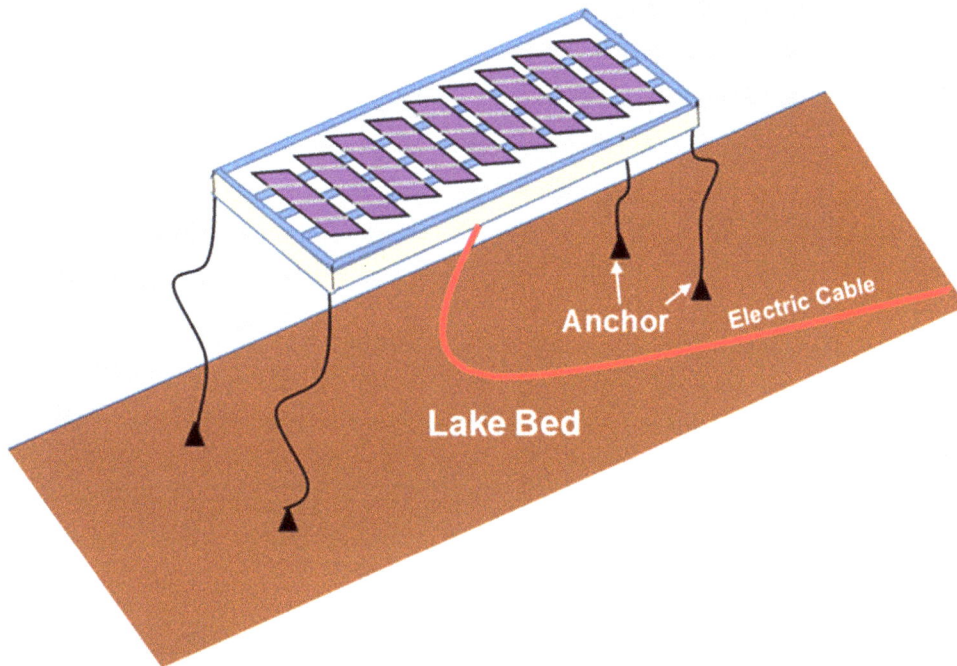

Fig. 88

Asking the public where they would like to see electricity power supply systems, such as Wind Farms, many say "out in the sea", "anywhere but on land." Well, Solar Power like wind power, it can also be floating.

Floating Solar Farms (Fig. 88) is not exactly a Physical Mechanical Generator, but the Floating Solar Farm does produce electricity. China, India, Portugal and South Korea have some of the largest Floating Solar Power Farms in the world.

Ideal locations for sea panels would be in calmer equatorial seas, however offshore Solar Farms do present all sorts of challenges, although it is not out of the question. One of the main challenges is the unpredictability of the tide and waves is a problem if they were installed on the sea. which is why they prefer installing them in lakes or reservoirs, where the water is a bit calmer and more predictable.

Anchors have to secure these units to the lake, reservoir or seabed, as even lake or reservoir water can become choppy.

Disadvantages are few, but being so large, cleaning the Solar Surfaces is a mammoth task especially if located in the sea where the problem of salt also arises. Floating Farms would also be expensive to build and maintain. As mentioned, Solar energy is a Renewable energy.

Index:

www.ingramcontent.com/pod-product-compliance
Lightning Source LLC
Chambersburg PA
CBHW041621220326
41597CB00035BA/6188